BUSINESS ISSUES, COMPETITION AND ENTREPRENEURSHIP

KNOWLEDGE CREATION AND TRANSFER

NEW RESEARCH

BUSINESS ISSUES, COMPETITION AND ENTREPRENEURSHIP

Additional books in this series can be found on Nova's website under the Series tab.

Additional e-books in this series can be found on Nova's website under the e-book tab.

BUSINESS ISSUES, COMPETITION AND ENTREPRENEURSHIP

KNOWLEDGE CREATION AND TRANSFER

NEW RESEARCH

EDWIN MATOS
EDITOR

nova publishers
New York

NOTICE TO THE READER

Library of Congress Cataloging-in-Publication Data

ISBN: 978-1-63485-656-0

Published by Nova Science Publishers, Inc. † New York

CONTENTS

PREFACE

This book provides new research on knowledge creation and transfer. The first chapter of the book reviews social metacognition and knowledge creation in groups of people. Chapter Two presents an Action Research (AR) study during an organizational innovation process in a Portuguese company. Chapter Three presents a review of the literature on the involvement of clinical teams and patients-as-partners in the changes made to clinical practices. Chapter Four helps the readers better understand the role of internal stakeholders in developing countries to offer potential insights into the cross-border transfer of knowledge.

Chapter 1 - While a single person can have a creative insight, groups of people can work together to create new, useful ideas (knowledge creation) that they cannot create alone. Some group characteristics and processes (diversity, argumentation, politeness, shared positioning) facilitate knowledge creation. Groups with diverse views can create more ideas, representations, justifications, and solution proposals, especially when group members understand and value one another's diverse contributions. During argumentation, groupmates explain and justify their own views to convince both themselves and others. During this process, group members evaluate one another's ideas and recognize problems. In contrast, other group processes (public self-image [face] and status concerns) hinder it. Although polite disagreements might aid knowledge creation, rude disagreements can reduce the likelihood of knowledge creation, especially when individuals are concerned about status. Also, group members might prefer to protect their face rather than contribute to knowledge creation. In general, group members' monitoring (e.g., evaluations) and control (e.g., commands) of one another's knowledge, emotions and actions (social metacognition) can influence group knowledge creation.

Chapter 2 - This chapter presents an Action Research (AR) study during an organizational innovation process in a Portuguese company. The central question is: do organizational actors share their individual knowledge to help the company in the organizational innovation process?

In order to accomplish that goal the authors used AR methodology to analyse if the employees share their knowledge and effectively integrate it in organization practices and processes.

A main result of this research is that knowledge sharing processes became part of the strategy of innovation of the company for the last decade and to implement it they have developed a culture of knowledge and mechanisms that facilitate the sharing and the integration of knowledge in day-by-day work practices.

The methodology used in the organizational innovation project helped to create an involvement and participation culture, making them more aware of the importance of new organizational practices and processes.

Chapter 3 - Despite numerous Quebec initiatives to improve transitional care for patients with chronic diseases such as cancer, clinical teams and patients still encounter some difficulties during these transitions. This article presents a review of the literature on the involvement of clinical teams and patient-as-partners in the changes made to clinical practices. It also highlights the main issues that arise during the implementation of a knowledge exchange strategy with a *Forum for Knowledge Exchange* (FKE) that aims to guide clinical teams and patients to improve the continuum of care in oncology. The FKE is an interactive space for researchers, nurses, other healthcare professionals, and patient-as-partners where they can share their knowledge, both tacit and scientific. This study uses a participative methodology and is conducted in six clinical settings that span the continuum of care in oncology. Patients-as-partners are an integral part of the implementation process, as active participants rather than in a simply advisory role. This strategy should help improve transitions between the different phases of the care continuum. The co-construction of knowledge stemming from this clinical research study should lead to a renewal of clinical practices and thus guide clinical decision-making in multidisciplinary teams and patients.

Chapter 4 - The purpose of this chapter is to better understand the role of internal stakeholders in a developing country to offer potential insights into the cross-border transfer of knowledge. The study involves case studies of four companies that invest in Mozambique in business areas which are presently of interest. The results obtained reveal insights that linked successful knowledge transfer to overseas with a set of operational factors: different perception of local internal stakeholders of what is happening in organizational context; affirmation of the working philosophy of the parent company underlying the organizational practices transferred; operational relevance of the transferred knowledge; and adaptation of the practices to the local context. The case study method does not allow the generalization of the results, but it permits the study of a combination of problems of the phenomenon that contribute to its better understanding. This chapter contributes to multinational corporations to know how to transfer knowledge to their overseas subsidiaries that operate in the industrial sectors, taking into consideration the particular characteristics of the market and the absorptive capacity existing in lesser developed countries.

In: Knowledge Creation and Transfer
Editor: Edwin Matos
ISBN: 978-1-63485-656-0

Chapter 1

SOCIAL METACOGNITION AND KNOWLEDGE CREATION IN GROUPS

*Ming Ming Chiu**
Purdue University, West Lafayette, IN, US

ABSTRACT

While a single person can have a creative insight, groups of people can work together to create new, useful ideas (knowledge creation) that they cannot create alone. Some group characteristics and processes (diversity, argumentation, politeness, shared positioning) facilitate knowledge creation. Groups with diverse views can create more ideas, representations, justifications, and solution proposals, especially when group members understand and value one another's diverse contributions. During argumentation, groupmates explain and justify their own views to convince both themselves and others. During this process, group members evaluate one another's ideas and recognize problems. In contrast, other group processes (public self-image [face] and status concerns) hinder it. Although polite disagreements might aid knowledge creation, rude disagreements can reduce the likelihood of knowledge creation, especially when individuals are concerned about status. Also, group members might prefer to protect their face rather than contribute to knowledge creation. In general, group members' monitoring (e.g., evaluations) and control (e.g., commands) of one another's knowledge,

* Educational Psychology and Research Methodology, Purdue University, 5156 Beering Hall, 100 North Street, West Lafayette, IN, 47907, US. Email: mingmingchiu@gmail.com.

emotions and actions (social metacognition) can influence group knowledge creation.

Introduction

Past research on the development of original ideas that are useful or influential (creativity) has largely focused on individuals (Runco 2014). However, the explosion of information and specializations will increasingly require teams with diverse skills and knowledge to create innovations (group creativity; Sawyer 2014). For example, groups are more likely than the best individual members to solve some problems (e.g., letters-to-numbers problems, Laughlin et al. 2006). Recently, more researchers have studied group creativity in organizations (Sawyer 2014) and collaborative learning (Johnson, Johnson, and Smith 2007). However, researchers have not systematically examined how recent group processes (micro-time context) affect the likelihood of creative moments (group knowledge creation).

A brief discussion of terminology might be helpful. *Creativity* is the generation of new ideas that are useful, and it can occur at different levels: for an entire society, for a single individual or at a moment in time (Runco 2014). Big "C" creativity affects many people in a society (e.g., smart phones, continuous improvement process, and so on) (Gruber and Wallace 1999). In contrast, small "c" creativity may affect only a single person (e.g., buying a coffee maker to avoid the morning traffic at a local coffee shop) (Gruber and Wallace 1999). Meanwhile, *knowledge creation* occurs when a person creates a useful, new idea at a specific moment in time (e.g., combining two colleagues' suggestions to create a new idea; Chiu 2008a).

Diversity

Groups with diverse views can create more ideas, representations, justifications, and solution proposals, especially when group members value one another's diverse contributions (Larson 2007; Paulus and Brown 2003; Swann et al. 2003). Specifically, group diversity in nationality or ethnicity often increases the number of perspectives, number of ideas, and quality of ideas (Paulus and Brown 2003). Group members' diverse views can also help them justify the validity of an idea, identify flaws, and revise them to create new ideas (Orlitzky and Hirokawa 2001). Hence, group diversity can enhance knowledge creation.

Group members with diverse views can also build on one another's ideas through processes such as sparked ideas, jigsaw pieces, and creative misinterpretations (Paulus and Brown 2003). Comments by one person (e.g., a key word) might spark another person to activate related concepts in his or her semantic network and propose a new idea (Nijstad, Diehl, and Stroebe 2003). Like fitting jigsaw pieces together, group members also can put together ideas to create a new idea (Milliken, Bartel, and Kurtzberg 2003). Lastly, a person might also misinterpret another person's incorrect idea to create a new idea (Chiu 2008a). Thus, even wrong ideas can be kindling for new ideas. This contrasts with the view that people only build on correct ideas and that wrong ideas lead the group astray. Hence, both correct and incorrect, new ideas might aid knowledge creation.

METACOGNITION AND SOCIAL METACOGNITION

Individual *metacognition* is monitoring and controlling one's own knowledge, emotions, and actions (Hacker and Bol 2004). For example, Hector believes he has set his alarm clock properly for an important business meeting in the morning, but he is not sure, so he checks his alarm again. Hector monitors his knowledge of the alarm setting, recognizes his lack of confidence in his knowledge and thus, directs himself to act to collect further information about it.

Social metacognition is the social analog to individual metacognition (Chiu and Kuo 2009). For example, a Chief Executive Officer (CEO) often distributes responsibilities among different officers, such as a Chief Financial Operator (CFO) who oversees the firm's finances, a Chief Operating Officer (COO) who oversees the firm's day-to-day operations, a Chief Technical Officer (CTO) who oversees the firm's technology and so on. After a catastrophe such as an earthquake, the CEO meets with senior staff to hear their assessments of the earthquake's impact on the firm, asks for options for rebuilding the company and then for evaluations of these options. The CEO listens to senior staff's ideas, evaluates them, and decides on which actions to take.

A Conversation Illustrating Social Metacognition's Effects on Knowledge Creation

Group members' social metacognition can enhance or hinder a person's knowledge creation (Chiu 2008a), as illustrated in the following annotated scenario.

> Xiao works at Local Expertise, an onsite consulting firm catering to the needs of travelers. He receives a memo to gather some colleagues and develop ideas to improve their firm's marketing. Xiao asks his good friend Diego and their co-worker Elena, who often has great ideas, to work together on this initiative.

Xiao uses his social metacognitive knowledge of his colleagues to select appropriate team members for this task, namely employees who understand one another, work together comfortably and generate interesting ideas (Sawyer 2014).

Status

Working with co-workers reduces status effects, which can hinder free exchanges of ideas and distort evaluations toward higher status colleagues. Status differences can hinder collaborative problem solving through the greater influence of high status members (Dembo and McAuliffe 1987). Cohen (1994) defined status as "an agreed-on rank order where it is generally felt to be better to be high than low rank" (p. 23).

Status affects the expectations of individual group members and is linked to the expectation of competencies for the current activity (*expectation states* theory, Berger, Cohen, and Zelditch 1972). High status is conferred on group members who are expected to contribute positively to a desired outcome. These expectations create different opportunities to perform and receive rewards. Members can selectively invite and defer to high status members' opinions while discouraging, undervaluing, or outright ignoring lower status members' ideas. By doing so, members enact their expectations of high status members dominating the interaction to the detriment of lower status members during the initial stages of group activity (Turner 2005).

Domination by higher status group members in early group interactions creates initial group member identities and group discourse norms.

Afterwards, group members often prefer and seek self-verifying evaluations of their initial identities from others (Swann 2005, Swann et al. 2003). Furthermore, the patterns of early discourse often become the preferred discourse norms (Sfard 2007). As group members value and prefer supporting previously discussed or shared information rather than introducing unshared information, high status members' influence further increases in over time (Chiu and Khoo 2003). Hence, lower status group members might enact these lower expectations and identities, withhold their ideas, and reduce the group's overall knowledge creation. Thus, greater status differences increases greater status effects, which reduces knowledge creation.

> Next, Diego and Elena agree to meet with Xiao.
>
> Diego: We already have our own website, isn't that enough?
>
> Elena: Yes, a website is essential but we need to get people to go there. Maybe we can use some social networking to get them there?
>
> Xiao: Good idea, what if we started providing online coupons by using Connected Fun?

Sharing relevant information by citing their current website, Diego criticizes their current task goal by questioning its necessity ("isn't that enough?"), thereby inviting and motivating Elena's knowledge creation to address his criticism. Hence, this excerpt exemplifies argumentation in the cognitive/problem content space, a social process by which people explain and justify their own views to convince both themselves and others (Jehn and Mannix 2001). During this process, group members evaluate one another's ideas, recognize problems, and justify their views (Chiu 2008a).

Diego's negative evaluation tries to change the current trajectory of the group. In general, agreements encourage continuation of the current problem-solving trajectory, while disagreements and changes of topic (ignoring the previous action) try to change it (West 2000). Evaluations characterize how a person assesses the previous speaker's action and problem solving approach (*functional theory of group decision-making*, Orlitzky and Hirokawa 2001).

Diego's disagreements might increase knowledge creation both directly and indirectly (*socio-cognitive conflict* theory, Buchs et al. 2004). A group's collective attention and diverse perspectives can help it identify more flaws compared to individuals (Milliken, Bartel, and Kurtzberg 2003). By identifying an obstacle, Diego's disagreement might directly stimulate knowledge creation, especially from passive, social loafing students who no longer rely on other group members to solve the problem. Furthermore, a disagreement (even if wrong) often stimulates the attention of group members (De Dreu and West 2001). Thus, a disagreement might help them consider more aspects of problem from more perspectives to increase knowledge creation indirectly.

Furthermore, disagreements might generally encourage group members with minority views to express their ideas, especially after agreements and repetitions of an existing idea suggest a majority view (Nemeth and Chiles 1988). Thus, Diego's disagreement, regardless of its validity, legitimizes different opinions, freeing all group members to express their ideas, including those unrelated to the specific disagreement (Nemeth and Chiles 1988). This greater legitimacy of different opinions can aid knowledge creation.

Moreover, Diego does not present his negative evaluation as definitive. Instead, he invites his colleagues' views by asking a question ("isn't that enough?"). (To aid reader comprehension, "question," "command," and "statement" are used rather than "elicitation," "directive," and "informative" [Parker 2014].) A person asking a question invites others to share information (Parker 2014).

Elena agrees with Diego's idea and validates it ("yes, a website is essential") rather than immediately disagreeing with him, showing her social metacognitive understanding of his emotions and his thinking. She listens to his criticism and understands it. By agreeing with him, she promotes his public self-image (gives *face*), provides emotional support and builds social rapport with him (Chiu 2008a). By repeating their shared information ("website"), Elena shows shared knowledge, common ground and solidarity (Stasser and Birchmeier 2003). Together, her agreement and repetition encourages him to continue listening to her idea (Chiu and Khoo 2003), which increases their group efficiency and effectiveness (e.g., Civettini 2007).

However, desire for agreement might reduce expression of new ideas. A group member with both supportive information and conflicting information tends to agree and provide supportive information, which is likely more salient and hence more likely to be activated quickly (Larson et al. 1996). Even if both pieces of information are activated, the common information tends to be more credible and compelling because socially validated consensus is often correct (Chiu 2008a). Based on probabilities, a common idea shared by many people is more likely to be voiced and supported than an unusual idea held by a few people (Stasser and Birchmeier 2003). Hence, agreements support face but can reduce expression of new ideas.

Instead of simply agreeing with Diego, Elena socially positions their task responsibility as shared ("we need to") and identifies the inadequacy of the website ("but we need to get people to go there"), so her social metacognitive action validates the task goal and deters Diego's attempt to close the discussion (Chiu and Khoo 2003). Furthermore, specifying the inadequacy motivates the group's need for new ideas (*knowledge creation*), which she addresses by sharing her new idea ("social networking," Chiu 2008a). Showing social metacognitive understanding of her colleagues' emotions, thinking and actions, she socially positions (Wardhaugh and Fuller 2014) her idea as belonging to the group ("we can") and engages Xiao and Diego to consider it by asking them to evaluate its usefulness ("maybe …?").

Xiao evaluates Elena's idea positively ("Good idea"), which gives face to Elena, enhances their social relationship and helps build shared knowledge. Then, he uses their shared knowledge and builds on it by suggesting a specific implementation ("providing online coupons by using Connected Fun"), showing how other's knowledge creation can spark one's own knowledge creation. Next, Elena asks about "Connected Fun."

> Elena: Connected Fun?
>
> Xiao: Connected Fun emails people and invites them and their friends to buy online discount coupons.
>
> Elena: That's a great idea.
>
> Not familiar with Xiao's idea, Elena shares her self-monitoring with the group and asks for more information ("Connected Fun?"). Xiao explains his new idea to help Elena understand it, support it and to build shared knowledge within the group. While Elena supports Xiao's idea, Diego does not.
>
> Diego: Well, like most people, I am not a fan of online shopping because putting personal information online is too risky.
>
> Elena: Actually, there are plenty of ways consumers can shop online securely, like with PayPal.
>
> Diego: That doesn't change my mind… I still wouldn't do it.
>
> Elena: I'm sure there are people that feel that way too, but we can focus on the consumers that do shop online.

Diego identifies a potential problem ("personal information online is too risky"), but shows poor social metacognitive knowledge of others by incorrectly claiming that most people share his view ("like most people, I am not a fan"). Whereas Diego criticized the task earlier, here he criticizes Xiao's idea and threatens his face.

Unlike her earlier, diplomatic response about the goal of the task, Elena defends Xiao's idea (Paulus 2000) by rudely rejecting Diego's flawed idea as disconnected from reality ("actually"). Although Elena correctly evaluates Diego's incorrect claim, she reveals her inadequate social metacognitive control through her rude rejection, which threatens Diego's face (*face attack*, Tracy 2008), and reduces his psychological safety, willingness to listen and knowledge creation (Turner and Horvitz 2001). Moreover, she escalates the interpersonal conflict (Chiu 2008a). In the worst case, a spiral of rude disagreements can kill the collaboration. Even if the collaboration survives after a rude disagreement (or some other rude action, e.g., insult), group members might withhold their ideas rather than risk losing face (Mulryan 1992). Elena continues with a statement of a counter-claim ("consumers can shop online securely") and justifies it with evidence ("like with PayPal").

After Elena's rude rejection, Diego also responds rudely by declaring Elena's information as inadequate ("that doesn't change my mind") and maintaining his personal position ("I still wouldn't do it"). By focusing on himself ("I") as a data point to legitimize his own ideas and position himself as a key authority, Diego separates himself from the group. By doing so, he threatens the group's shared responsibility, shared positioning and cohesion, which hinders their cooperation and the group's knowledge creation.

Rather than accelerating the spiral of rude disagreements, Elena tries to reduce the tension by politely agreeing that some people share Diego's concern ("I'm sure there are people that feel that way too"), by re-establishing their shared positioning ("we can") and compartmentalizing the task ("focus on the consumers that do shop online"). At this point, Xiao intervenes.

> Xiao: Let's do some research on how Connected Fun works before making a decision. I can check on contact information.
> Elena: Great, I can look at coupon characteristics.
> Xiao: Diego, do you want to look at Connected Fun reviews?
> Diego: Sure, that sounds good.

Xiao uses his social metacognitive understanding of the tension between Elena and Diego to suggest collecting more information ("let's do some research on how Connected Fun works") and postpone the decision ("before making a decision"), which adjourns the discussion and stops the argument that had stunted their knowledge creation. By having Diego and Elena gather more information, Xiao helps them become more comfortable with Connected Fun, acquire more ideas and develop more shared knowledge, all of which can stimulate their knowledge creation. If Xiao had let Elena and Diego's disagreement fester, it could have threatened their faces, damaged their social relationship, and hindered their knowledge creation.

Next, Xiao and Elena volunteer to gather specific information ("contact information," "coupon characteristics"). When Xiao politely asks Diego to consider collecting other information ("do you want to look at Connected Fun reviews?"), Diego agrees. As Diego's face has already been threatened by Elena's face attack, he is more likely to agree to a request from Xiao (especially a polite one) rather than Elena.

> Everyone in the group starts an internet search on their laptops for Connected Fun in hopes to find some useful information.
>
> Diego: I can't believe it. It's supposed to rain all weekend. Great, there go my plans.
>
> Xiao: The weather forecast is always changing. Check later and it will be mostly sunny.
>
> Elena: Come on, guys. Let's worry about the weekend after our meeting.

Diego gets distracted by a weather forecast ("it's supposed to rain"), which distracts Xiao as well ("weather forecast is always changing"). Monitoring her co-workers' attention, Elena refocuses their attention on the task ("let's worry about the weekend after our meeting") to reduce further distractions and enhance their knowledge creation.

Social Metacognition Effects on Knowledge Creation

The above conversation illustrates how social metacognition via understanding others, evaluations, positioning and questions, can aid knowledge creation (see Table 1). Understanding others through monitoring can aid in selecting appropriate team members who have the necessary talent and who can work together to create new ideas to accomplish a task. As a team works on a task, monitoring enables team members to recognize entangling distractions, tensions among group members, and differences in their views. After recognizing that others are distracted, a person can re-direct their attention to the task to enhance knowledge creation.

Upon recognizing rising tension among team members, a person can try to defuse the tension by highlighting/building agreement among them or by suggesting a different activity. When team members' views differ, understanding how to capitalize on the differences can aid knowledge creation, while being frustrated by them hinders knowledge creation.

Evaluations can facilitate knowledge creation through criticisms and agreements, especially correct evaluations. Criticisms identify potential flaws and motivate knowledge creation to address them (or counter-arguments and justifications to reject the flaws). However, rude disagreements (such as face attacks) can escalate interpersonal conflict, and hinder knowledge creation. As politeness during disagreement is the norm, lack of redress is noticeable and considered impolite (Holtgraves 1997). Hence, criticisms are often rendered polite by accompanying redress such as specifying partial agreements, highlighting shared positioning and inviting further evaluation through questions.

Meanwhile, agreements can help build emotional and cognitive foundations for knowledge creation. Agreeing with the perceived correct parts of others' ideas gives/saves face and supports a shared, secure emotional foundation. Emotionally secure group members can listen to and explore a wider range of ideas without fearing personal rejection or embarrassment. Eventually, they can accept rejections of their weaker ideas in favor of others' stronger ideas. Group members' agreements also help build a cognitive foundation of shared understanding of greater knowledge that stimulates their exploration, development and evaluation of new ideas. The validity of an evaluation can also affect knowledge creation. Correct evaluations support correct ideas or identify flaws to support the group's shared, correct understandings, which can launch micro-creative elaborations. In contrast, incorrect evaluations reject correct ideas or accept flawed ideas, resulting in poorer, shared understandings that can stimulate wrong, new ideas and less knowledge creation.

Positioning (Davies and Harre 1990) task responsibilities and ideas as shared among group members is a form of social metacognitive control that helps build a social foundation for subsequent knowledge creation (Chiu and Kuo 2009). Shared positioning can share risk by distributing responsibilities among group members and aid emotional support to enhance motivation. As group members share responsibilities, they also share the risks of failure and its consequences. With less personal risk and a lower cost of failure, collaborators can feel less anxious and more motivated to create new ideas. In contrast, positioning oneself in opposition to others (I vs. you) heightens one's sense of authority based on one's own experiences, separates oneself from the group, reduces group cohesion, reduces sharing of ideas and hinders knowledge creation.

Table 1. Mechanisms by which Social Metacognition affects Knowledge Creation

Social Metacognition	→ Mechanism → Knowledge creation
Understand/ Monitor colleagues	→ Select appropriate group members with suitable skills who work well together → Knowledge creation → Monitor tension → Reduce tension → Redirect attention to task → Knowledge creation → Monitor their attention → Identify distractions → Return their attention to task → Knowledge creation
Disagree/ Criticize politely	→ Detect flaws → Motivate correction of flaws → Knowledge creation → Give/save face → Emotional security → Explore ideas → Knowledge creation
Disagree/ Criticize rudely	→ Face attack → Reduces other's openness to listen → Invites retaliation → Deadly spiral of face attacks –X→ Knowledge creation → More tension within group → Decreases group cooperation → Promotes individual positioning (I) → Reduces ideas explored –X→ Knowledge creation
Agree/Support	→ Support ideas → Give/save face → Emotional security → Explore ideas → Knowledge creation
	→ Support other's ideas → Develop shared foundation of greater knowledge → Stimulate related ideas → Knowledge creation

Table 1. (Continued)

Social Metacognition	→ Mechanism → Knowledge creation
Evaluate correctly	→ Support correct ideas + Detect flaws → Develop shared foundation of correct knowledge → Stimulate related ideas → Knowledge creation
Evaluate incorrectly	→ Discourage correct ideas + Support flaws → Develop shared foundation of flawed knowledge → Stimulate flawed ideas –X→ Knowledge creation
Social Metacognition	→ Mechanism → Knowledge creation
Shared positioning (we)	→ Shared responsibility → Shared ideas → Knowledge creation → Shared risk of failure → Lower individual risk → Less anxiety → Greater motivation → Explore ideas → Knowledge creation
Oppositional positioning (I vs. you)	→ Authority based on own experience → Separate responsibility → Individual risk of failure → Less sharing of ideas –X→ Knowledge creation
Ask for evaluation	→ Invite attention, consideration and evaluation by others → Develop shared foundation of greater knowledge → Elaborate idea → Knowledge creation → Detect flaw → Motivate correction of flaw → Knowledge creation
Ask for information	→ Solicit information → Group member(s) fill knowledge gap → Develop shared foundation of greater knowledge → Stimulate related ideas → Knowledge creation
Command	→ Reusing old ideas –X→ Knowledge creation → Harm social relationships → Exacerbate status effects → Reduce ideas explored –X→ Knowledge creation

Note: –X→ indicates "hinders."

Questions are a form of social metacognitive control that invites evaluations or solicits information to support a cognitive foundation for subsequent knowledge creation (Chiu 2008b). By framing ideas in the form of questions, a person invites group members to consider an idea, evaluate its validity and usefulness, and create new ideas that elaborate it or address its flaws. Unlike ideas expressed as definitive statements or commands, ideas in the form of questions are more polite, so they are less likely to draw a negative response, prematurely truncate the discussion or discourage identification of weaknesses. Questions that solicit information invite other group members to fill the gap in the group's shared cognitive foundation, which might otherwise hinder exploration, development and evaluation of new ideas.

CONCLUSION AND FUTURE DIRECTIONS

Micro-analyses of conversations can show how colleagues can influence one another, specifically how social metacognition (via understanding others, evaluations, positioning and questions) can aid the creation of new ideas (knowledge creation). Research on social metacognition and knowledge creation is in its infancy with many unanswered theoretical and methodological questions. Theoretical issues include the relationships of knowledge creation to the small "c" creativity of individuals and groups and the Big "C" creativity of society. Furthermore, the relationships between the above social metacognitive actions and knowledge creation might differ along groups and its member characteristics (demographics, individual histories, interpersonal relationships, distribution of relevant competences), contexts (activity, institution, industry, culture) and periods of time (hours, weeks, months, years). In addition to the insights offered by case studies, methodology developments in statistical discourse analysis (Chiu 2008a) are needed to systematically study many conversations by many groups. Still, social metacognition's effects on knowledge creation show how team members' interactions can affect their team creativity, and further research in this area might yield interventions that enhance team creativity.

References

Berger, Joseph., Bernard P. Cohen, and Morris Zelditch Jr. 1972. "Status Characteristics and Social Interaction." *American Sociological Review 37*: 241–56.

Buchs, Celine, Fabrizio Butera, Gabriel Mugny, and Celine Darnon. 2004. "Conflict Elaboration and Cognitive Outcomes." *Theory into Practice 43*: 23–30.

Chiu, Ming Ming, and Lawrence Khoo. 2003. "Rudeness and Status Effects During Group Problem Solving." *Journal of Educational Psychology 95*: 506–23.

Chiu, Ming Ming, and Sze Wing Kuo. 2009. "From Metacognition to Social Metacognition: Similarities, Differences, and Learning." Education Research Journal 3: 1–19.

Chiu, Ming Ming. 2008a. "Flowing Toward Correct Contributions During Group Problem Solving: A Statistical Discourse Analysis." *Journal of the Learning Sciences 17*: 415–63.

Chiu, Ming Ming. 2008b. "Effects of Argumentation on Group Knowledge Creation." *Contemporary Educational Psychology 33*: 383–402.

Civettini, Nicole H. W. 2007. "Similarity and Group Performance." *Social Psychology Quarterly 70*: 262–71.

Cohen, Elizabeth G. 1994. "Restructuring the Classroom: Conditions for Productive Small Groups", *Review of Educational Research. 64*: 1–35.

Davies, Brownyn. and Rom Harre. 1990. "Positioning: The Discursive Production of Selves." *Journal for the Theory of Social Behavior 20*: 43–63.

De Dreu, Carsten K. W., and Michael A. West. 2001. "Minority Dissent and Team Innovation." *Journal of Applied Psychology 86*: 1191–201.

Dembo, Myron H., and Timothy J. McAuliffe. 1987, "Effects of Perceived Ability and Grade Status on Social Interaction and Influence in Cooperative Groups." *Journal of Educational Psychology 79*: 415–23.

Gruber, Howard E., and Doris B. Wallace. 1999. "The Case Study Method and Evolving Systems Approach for Understanding Unique Creative People at Work." In *Handbook of creativity*, edited by R. J. Sternberg, 93–115. Cambridge, UK: Cambridge University Press.

Hacker, Douglas J., and Linda Bol. 2004. "Metacognitive Theory: Considering the Social-cognitive Influences." In *Big Theories Revisited,* edited by Dennis M. McInerney and Shawn Van Etten, 275-97. Greenwhich, Connecticut: Information Age Publishing, Inc.

Holtgraves, Thomas. 1997. "YES, BUT… Positive Politeness in Conversation Arguments." *Journal of Language and Social Psychology 16*: 222–39.

Jehn, Karen A., and Elizabeth A. Mannix. 2001. "The Dynamic Nature of Conflict." *Academy of Management Journal 44*: 238–51.

Johnson, David W., Roger T. Johnson, and Karl Smith. 2007. "The State of Cooperative Learning in Postsecondary and Professional Settings." *Educational Psychology Review 19*: 15–29.

Larson, James R., Jr. 2007. "Deep Diversity and Strong Synergy." *Small Group Research 38*: 413–36.

Larson, James R., Jr., Caryn Christensen, Timothy M. Franz, and Ann S. Abbott. 1998. "Diagnosing Groups: The Pooling, Management, and Impact of Shared and Unshared Case Information in Team-based Medical Decision Making." *Journal of Personality and Social Psychology* 75: 93–108.

Laughlin, Patrick. R., Erin C. Hatch, Jonathan S. Silver, and Lee Boh. 2006. "Groups Perform Better Than the Best Individuals on Letters-to-Numbers Problems." *Journal of Personality and Social Psychology 90*: 644–51.

Milliken, Frances J., Caroline A. Bartel, and Terri R. Kurtzberg. 2003. "Diversity and Creativity in Work Groups." In *Group Creativity: Innovation Through Collaboration,* edited by P. B. Paulus and B. A. Nijstad, 32–62. New York: Oxford University Press.

Mulryan, Catherine M. 1992. "Student Passivity During Cooperative Small Groups in Mathematics." *Journal of Educational Research 85*: 261–73.

Nemeth, Charlan, and Cynthia Chiles. 1988. "Modeling Courage: The Role of Dissent in Fostering Independence." *European Journal of Psychology 18*: 275–80.

Nijstad, Bernard A., Michael Diehl, and Wolfgang Stroebe. 2003. "Cognitive Stimulation and Interference in Idea Generating Groups." In *Group Creativity: Innovation Through Collaboration,* edited by Paul B. Paulus and Bernard A. Nijstad, 137–59. New York: Oxford University Press.

Orlitzky, Marc, and Randy Y. Hirokawa. 2001. "To Err Is Human, to Correct for It Divine." *Small Group Research 32*: 313–41.

Parker, Ian. 2014. *Discourse Dynamics: Critical Analysis for Social and Individual Psychology*. London: Routledge.

Paulus, Paul B., and Vincent R. Brown. 2003. "Ideational Creativity in Groups." In *Group Creativity: Innovation Through Collaboration,* edited by Paul B. Paulus and Bernard A. Nijstad, 110–36. New York: Oxford University Press.

Paulus, Paul B. 2000. "Groups, Teams, and Creativity." *Applied Psychology 49*: 237–62.

Runco, Mark A. 2014. *Creativity: Theories and Themes: Research, Development, and Practice*. London: Academic Press.

Sawyer, R. Keith. 2014. *Group creativity: Music, Theater, Collaboration*. London: Psychology Press.

Sfard, Anna. 2007. "When the Rules of Discourse Change, But Nobody Tells You." *The Journal of the Learning Sciences 16*: 567–615.

Stasser, Garold, and Zachary Birchmeier. 2003. "Group Creativity and Collective Choice." In *Group Creativity: Innovation Through Collaboration,* edited by Paul B. Paulus and Bernard A. Nijstad, 85–109. New York: Oxford University Press.

Swann, William B., Jr. 2005. "The Self and Identity Negotiation." *Interaction Studies 6*: 69–83.

Swann, William B., Jr., Virginia S. Y. Kwan, Jeffrey T. Polzer, and Laurie P. Milton. 2003. "Fostering Group Identification and Creativity in Diverse Groups." *Personality and Social Psychology Bulletin 29*: 1396–406.

Tracy, Karen. 2008. "Reasonable Hostility." *Journal of Politeness Research 4*: 169–91.

Turner, John C. 2005. "Explaining the Nature of Power: A Three-process Theory." *European Journal of Social Psychology 35*: 1–22.

Turner, John C., and Todd Horvitz. 2001. "The Dilemma of Threat: Group Effectiveness and Ineffectiveness under Adversity." In *Groups at Work: Theory and Research,* edited by Marlene E. Turner, 445–70. Mahwah, NJ: Erlbaum.

Wardhaugh, Ronald, and Janet M. Fuller. 2014. *An Introduction to Sociolinguistics*. New York: Wiley.

West, Michael A. 2000. "Reflexivity, Revolution, and Innovation in Work Teams." *Advances in Interdisciplinary Studies of Work Teams 5*: 1–29.

BIOGRAPHICAL SKETCH

Ming Ming CHIU
Purdue University

Education:

PhD, University of California, Berkeley, 1996
Education, Mathematics, Science and Technology

Ed. M., Harvard University, Interactive Technology, 1990
B. S., Columbia University, Computer Science, magna cum laude, 1988

Address: 5156 Beering Hall, Purdue University, 100 North Street, West Lafayette, IN, 47907

Research and Professional Experience:
Charles R. Hicks Professor of Educational Psychology and Research Methodology,
2014 – present
Dept of Educational Studies, Purdue University
Professor, Learning and Instruction 2008-2014
University at Buffalo, State University of New York
Associate Professor, Educational Psychology, 2002-2008; Assistant Professor, 1998-2002
The Chinese University of Hong Kong

Honors:
Outstanding Faculty Discovery Award, Purdue University,
Educational Studies, 2015
Top 50 Learning Sciences ideas: Statistical Discourse Analysis (SDA)
International Society of the Learning Sciences, 2013
Young Researcher, Chinese University of Hong Kong, 2005
Outstanding Teaching, Chinese University of Hong Kong
Faculty of Education, 1999, 2004, 2007
Leadership Award, Southwest Center, 1996 (US)

Publications Last 3 Years:

Book Chapters (36)

1. Chiu, M. M. (accepted). Statistical discourse analysis. In K. Peppler (Ed.) *SAGE Encyclopedia of Out-of-School Learning*. Thousand Oaks, CA: Sage.
2. Chiu, M. M. (accepted). Social metacognition and knowledge creation in groups. In N. Gotsiridze-Columbus's (Ed.) *Knowledge Creation: New Research*. Hauppauge, NY. Nova Science.
3. Chiu, M. M. (accepted). Chinese families. In Jenifer Kunz's (Ed.) *Family and Parent Leadership*. Thousand Oaks, CA: Sage.

4. Chiu, M. M. (accepted). Self-concept, self-efficacy and mathematics achievement: Students in 65 regions including the US and Asia. In J. Son, J. J. Lo, and T. Watanabe's (Eds.) *What matters? Research trends in international comparative studies in mathematics education in six countries*. New York: Springer.
5. Chiu, M. M. (accepted). Achievement tests. In J. Steinberg (Ed.) *Encyclopedia of Economics and Society*. Thousand Oaks, CA: Sage.
6. Chiu, M. M. (accepted). Asian-American families. In Jenifer Kunz's (Ed.) *Family and Parent Leadership*. Thousand Oaks, CA: Sage.
7. Chiu, M. M. (accepted). Socioeconomic status, inequality and academic achievement. In N. Gotsiridze-Columbus's (Ed.) *Socioeconomic Status: Influences, Disparities and Current Issues.* Hauppauge, NY. Nova Science.
8. Chiu, M. M. (accepted). Early childhood investments. In J. Steinberg (Ed.) *Encyclopedia of Economics and Society*. 4th edition. Thousand Oaks, CA: Sage.
9. Chen, G. and Chiu, M. M. (accepted). Discussion processes in online forums. In M. Khosrow-Pour's (Ed.) *Encyclopedia of Information Science and Technology* (pp. 2544-2554). Hershey, PA: IGI Global.
10. Chiu, M. M. (2015). Statistical discourse analysis. In L. Resnick, C. Asterhan and S. Clarke's (Eds.) *Socializing Intelligence through Academic Talk and Dialogue* (pp. 301-314). Washington, DC: American Educational Research Association [AERA].
11. Chiu, M. M. and Joh, S. W. (2015). Cross-national comparisons in education-PISA. In James D. Wright (Ed.) *International Encyclopedia of Social and Behavioral Sciences. 2nd Edition, Vol. 5* (pp. 342-348). Oxford: Elsevier.
12. Chiu, M. M. and Chow, B. W.-Y. (2015). International comparisons of student achievement. In S. Rosa's (Ed.) *Progress in Education, vol. 32* (pp. 93-108). Hauppauge, NY: Nova Science.
13. Arya, P., Christ, T. and Chiu, M. M. (2015). TAP (**T**eacher learning, **A**pplication to **P**edagogy) through video mediated discussions. In M. L. Niess and H. Gillow-Wiles's (Eds.) *Handbook of Research on Teacher Education in the Digital Age* (pp. 334-356). Hershey, PA: IGI Global.
14. Christ, T., Arya, P. and Chiu, M. M. (2015). A three-pronged approach to video reflection: Preparing literacy teachers of the future. In E. Ortlieb, L. Shanahan, and M. McVee (Eds.), *Video Reflection in*

Literacy Teacher Education and Development: Lessons from Research and Practice (pp. 235-253). Bingley, UK: Emerald.

15. Chiu, M. M. (2014). Family and school influences on students' math achievement: United States vs. China. In W. Ma's (Ed.) *East Meets West in Teacher Preparation: Crossing Chinese and American borders* (pp. 143-158). New York: Teachers College Press.
16. Chiu, M. M. and Fujita, N. (2014). Statistical discourse analysis of online discussions: Informal cognition, social metacognition and knowledge creation. In S.-C. Tan, H.-J. So and J. Yeo's (Eds.) *Knowledge creation in education* (pp. 97-112). New York: Springer.
17. Chen, G. and Chiu, M. M. (2014). Learning processes during online discussions. In M. Khosrow-Pour's (Ed.) *Encyclopedia of Information Science and Technology* (pp. 2544-2554). Hershey, PA: IGI Global.
18. Chiu, M. M. and Chen, G. (2014). Collectivism. In L. Ganong, M. Coleman, and G. J. Golson's (Eds.) *The Social History of the American Family* (pp. 254-256). Thousand Oaks, CA: Sage.
19. Chiu, M. M. and Chen, G. (2014). Individualism. In L. Ganong, M. Coleman, and G. J. Golson's (Eds.) *The Social History of the American Family* (pp. 721-724). Thousand Oaks, CA: Sage.
20. Chiu, M. M. and Pawlikowski, M. (2014). Head Start. In L. Ganong, M. Coleman, and G. J. Golson's (Eds.) *The Social History of the American Family* (pp. 660-662). Thousand Oaks, CA: Sage.
21. Chiu, M. M. (2013). Statistical discourse analysis of an online discussion: Cognition and social metacognition. In *Productive Multivocality in the Analysis of Collaborative Learning* (pp. 417-434). New York: Springer.
22. Chiu, M. M. (2013). Classroom interactions between students. In J. Ainsworth and G. J. Golson's (Eds.) *Sociology of Education* (pp. 125-127). Thousand Oaks, CA: Sage.
23. Chiu, M. M. (2013). Social metacognition, micro-creativity and justifications: Statistical discourse analysis of a mathematics classroom conversation. In *Productive Multivocality in the Analysis of Collaborative Learning* (pp. 141-160). New York: Springer.
24. Chiu, M. M. (2013). Class inequality –Achievement. In J. Ainsworth and G. J. Golson's (Eds.) *Sociology of Education* (pp. 116-118). Thousand Oaks, CA: Sage.
25. Chiu, M. M. and Chen, G. (2013). Statistical discourse analysis: Testing hypotheses with large datasets of electronic discourse. In H. L. Lim and F. Sudweeks's (Eds.) *Innovative Methods and*

Technologies for Electronic Discourse Analysis (pp. 285-303). Hershey, PA: IGI Global.

26. Chiu, M. M. and * Pawlikowski, M. J. (2013). Social metacognition and micro-creativity. In E. G. Carayannis's (Ed.), *Encyclopedia of Creativity, Invention, Innovation, and Entrepreneurship* (pp. 1687-1692). New York: Springer.
27. Chiu, M. M., * Molenaar, I. Chen, G., Wise, A. F. and Fujita, N. (2013). Micro-analysis of collaborative processes that facilitate productive online discussions: Statistical discourse analyses of three cases. In M. Clara and E. B. Gregori (Eds.) *Assessment and evaluation of time factors in online teaching and learning* (pp. 232-263). Hershey, PA: IGI Global.
28. Chiu, M. M. and Chow, B. W.-Y. (2013). International comparisons of student achievement. In M. Gowda and A. Khanderia's (Ed.) *Educational Achievement: Teaching Strategies, Psychological Factors and Economic Impact* (pp.115-132). Hauppauge, NY: Nova Science.
29. Chiu, M. M. (2013). Parental Involvement. In J. Ainsworth and G. J. Golson's (Eds.) *Sociology of Education* (pp. 570-572). Thousand Oaks, CA: Sage.
30. Chiu, M. M., * Jones, K. A. and * Jones, J. L. (2013). Building on Schoenfeld's studies of metacognitive control towards social metacognitive control. In Y. Li and J. Moschkovich's (Eds.) *Mathematical proficiency and beliefs in learning and teaching–learning from Alan Schoenfeld and Günter Toerner* (pp. 69-88). Rotterdam, Netherlands: Sense.
31. Chiu, M. M. (2013). Family structure and education. In J. Ainsworth and G. J. Golson's (Eds.) *Sociology of Education* (pp. 271-275). Thousand Oaks, CA: Sage.
32. Chiu, M. M. (2013). Grandparents' role in education. In J. Ainsworth and G. J. Golson's (Eds.) *Sociology of Education* (pp. 333-334). Thousand Oaks, CA: Sage.
33. Chiu, M. M. (2013). Adolescence. In J. Ainsworth and G. J. Golson's (Eds.) *Sociology of Education* (pp. 9-10). Thousand Oaks, CA: Sage.
34. Chiu, M. M. (2013). China. In J. Ainsworth and G. J. Golson's (Eds.) *Sociology of Education* (pp. 114-116). Thousand Oaks, CA: Sage.
35. Wise, A. F., Zhao, Y., Hausknecht, S. and Chiu, M. M. (2013). Temporal considerations in analyzing and designing for online discussions in education: Examining duration, sequence, pace and

salience. In M. Clara and E. B. Gregori (Eds.) *Assessment and evaluation of time factors in online teaching and learning* (pp. 198-231). Hershey, PA: IGI Global.

36. Chiu, M. M. (2013). Hong Kong. In J. Ainsworth and G. J. Golson's (Eds.) *Sociology of Education* (pp. 365-366). Thousand Oaks, CA: Sage.

Referred Journal Articles (27)

1. Chiu, M. M. and Roberts, C. A. (accepted). Improved analyses of single cases: Dynamic multilevel analysis. *Developmental Neurorehabilitation.*
2. Chiu, M. M. and Huang, X. (accepted). Statistical discourse analysis: Students' group problem solving. *Journal of Global Education.*
3. Reynolds, R. and Chiu, M. M. (accepted). Reducing digital divide effects through student engagement in coordinated game design, online resource uses, and social computing activities in school. *Journal of the American Society for Information and Technology.*
4. Lehmann-Willenbrock, N., Chiu, M. M., Lei, Z. and Kauffeld, S. (accepted). Understanding positivity within dynamic team interactions: A statistical discourse analysis. *Group Organization and Management.*
5. Bruce, D., Miller, S. M. and Chiu, M. M. (accepted). Rural students composing videos despite contextual constraints: Breaking down the in-school/out-of school divide. *Learning Media and Technology.*
6. Waight, N., Chiu, M. M. and * Whitford, M. M. (2014). Factors that influence science teachers' selection and usage of technologies in high school science classrooms. *Journal of Science Education and Technology, 23,* 5, 668-681.
7. Chiu, M. M., Chow, B. W. Y., McBride, C. and Mol, S. T. (2016). Students' sense of belonging at school in 41 countries: Cross-cultural variability. *Journal of Cross-cultural Psychology, 47,* 175-196.
8. Chiu, M. M. (2015). Family inequality, school inequalities and mathematics achievement in 65 countries: Microeconomic mechanisms of rent seeking and diminishing marginal returns. *Teacher's College Record, 117,* 1, 1-32.
9. Chiu, M. M. and Chow, B. W. Y. (2015). Classmate characteristics and student achievement in 33 countries: Classmates' past achievement, family SES, educational resources and attitudes toward reading. *Journal of Educational Psychology, 107,* 1, 152-169.

10. Hayden, H. E. and Chiu, M. M. (2015). Reflective teaching via a problem exploration-teaching adaptations-resolution cycle: A mixed methods study of pre-service teachers' reflective notes. *Journal of Mixed Methods Research, 9,* 2, *133-153.* doi:10.1177/1558689813509027
11. Bruce, D. L. and Chiu, M. M. (2015). Composing with new technology: Teachers reflecting on learning digital video *Journal of Teacher Education, 66,* 3, 272-287.
12. Cho, J. R. and Chiu, M. M. (2015). Rapid naming in relation to reading and writing in Korean (Hangul), Chinese (Hanja), and English among Korean children: A one-year longitudinal study. *Journal of Research in Reading, 38,* 4, 387-404.
13. Christ, T., Chiu, M. M. and Wang, X. C. (2015). Emergent readers' social interaction styles and their comprehension processes during buddy reading. *Literacy Research and Instruction, 54*, 1, 45-66.
14. Arya, P., Christ, T. and Chiu, M. M. (2015). Links between characteristics of collaborative peer video analysis events and literacy teachers' outcomes. *Journal of Technology and Teacher Education, 23*(2), 159-183.
15. Chiu, M. M. and Fujita, N. (2014). Statistical Discourse Analysis: A method for modelling online discussion processes. *Journal of Learning Analytics, 1* (3), 61–83.
16. Chiu, M. M., Joh, S. W. and Khoo, L. (2014). Effects of competition and exit threats on managerial incentives in the financial industry. *Review of Financial Information Studies, 3*, 71-89.
17. Molenaar, I. and Chiu, M. M. (2014). Dissecting sequences of regulation and cognition: Statistical discourse analysis of primary school children's collaborative learning. *Metacognition and Learning, 9*, 137-160.
18. Wise, A. F. and Chiu, M. M. (2014). The impact of rotating summarizing roles in online discussions: Effects on learners' listening behaviors during and subsequent to role assignment. *Computers in Human Behavior, 38,* 261-271.
19. Christ, T., Chiu, M. M. and Wang, X. C. (2014). Preschoolers' engagement with reading behaviours: A statistical discourse analysis of peer buddy-reading interactions. *Journal of Research in Reading, 37*, 4, 375-408.
20. Eryilmaz, E., Chiu, M. M., Thoms, B., Mary, J. and R. Kim. (2014). Design and evaluation of instructor-based and peer-oriented attention

guidance functionalities in an open source anchored discussion system, *Computers and Education, 71*, 303-321. doi: 10.1016/j.compedu.2013.08.009.

21. Christ, T., Arya, P. and Chiu, M. M. (2014). Teachers' reports of learning and application to pedagogy based on engagement in collaborative peer video analysis. *Teaching Education, 25,* 4, 349-374.
22. Arya, P., Christ, T. and Chiu, M. M. (2014). Facilitation and teacher behaviors: An analysis of literacy teachers' video-case discussions. *Journal of Teacher Education, 65* (2), 111-127.
23. Christ, T., Chiu, M. M., * Currie, A. and Cipielewski, J. (2014). The relation between test formats and kindergarteners' expressions of vocabulary knowledge. *Reading Psychology, 35,* 6, 499-528.
24. Reynolds, R. and Chiu, M. M. (2013). Formal vs. informal context factors as contributors to student engagement in a guided discovery-based program of game design learning. *Learning, Media and Technology, 38,* 4, 429-462.
25. Hayden, H. E. and Chiu, M. M. (2013). Lessons learned: Supporting the development of reflective practice and adaptive expertise. *Literacy Research Association Yearbook, 62*, 278-296.
26. Wang, X. C., Christ, T. and Chiu, M. (2013). Exploring a comprehensive model for early childhood vocabulary instruction: A design experiment. *Early Childhood Development and Care, 187*(7), 1075-1106. DOI: 10.1080/03004430.2013.843531
27. Moore-Russo, D., * Viglietti, J. M., Chiu, M. M. and * Bateman, S.M. (2013). Teachers' spatial literacy as visualizing, reasoning, and communication. *Teaching and Teacher Education, 29*, 97-109.

In: Knowledge Creation and Transfer
Editor: Edwin Matos

ISBN: 978-1-63485-656-0

Chapter 2

KNOWLEDGE CREATION THROUGH ACTION RESEARCH

Maria José Sousa*
CIEO – Algarve University, Faro, Portugal

ABSTRACT

This chapter presents an Action Research (AR) study during an organizational innovation process in a Portuguese company. The central question is: do organizational actors share their individual knowledge to help the company in the organizational innovation process?

In order to accomplish that goal we used AR methodology to analyse if the employees share their knowledge and effectively integrate it in organization practices and processes.

A main result of this research is that knowledge sharing processes became part of the strategy of innovation of the company for the last decade and to implement it they have developed a culture of knowledge and mechanisms that facilitate the sharing and the integration of knowledge in day-by-day work practices.

The methodology used in the organizational innovation project helped to create an involvement and participation culture, making them more aware of the importance of new organizational practices and processes.

Keywords: knowledge sharing, action research, organizations, innovation

* E-mail: mjsousa.pt@gmail.com.

INTRODUCTION

The research explicited in this chapter was developed in one innovative organisation from Portugal and specialized in robotics and automation. All workers participate in the process and share their experiences and knowledge in order to achieve high quality products. This research main goal was to analyse the perceptions of organizational actors regarding knowledge sharing trough the innovation processes in the organization activities and practices. The research was conducted in Efacec located in the north of Portugal and involved technicians and managers in separate group recall sessions where they shared experiences, ideas and gave suggestions about the knowledge sharing processes.

Five employees participated in each group recall session - totalizing 30 persons from the company. For data analysis we used analysis grids based on employee's quotations. We did not want to find out percentages or values and it was not our goal to make any kind of measures, but to understand the processes for knowledge sharing in the organization in innovation processes. In the field research, we will discuss forms of interaction in order to share tacit and explicit knowledge, supported by the base idea of Nonaka and Takeuchi's Knowledge Spiral. However, we will not use the categorisation of the model (Socialization, Externalization, Internalization, and Combination), because we think that the processes of creation and use/share of knowledge cannot be separated. It is a dynamic process that blends all forms of knowledge share.

This research was supported in the fact that individual knowledge is a resource that organizations can use to answer to challenges which the general environment is requiring, but its also important to analyse if they create and use mechanisms to generate and transfer new knowledge, not only regarding managers but also employees. The chapter is structured as follows: after the introduction, the literature review on knowledge sharing processes is reviewed, followed by the research methodology and the findings from the research.

THEORETICAL MODEL

Knowledge as been studied by different schools of thought: Organisational theory, Industrial Economy, Management, Innovation Management. All of them as analysed the form, results and processes through out organisations learn. The aim of this research is to analyse the role of individual knowledge in organisational innovation and change processes and in the literature we can find a strong linkage between them. However, knowledge can be an enabler or a disabler of organisational innovation and change success, because individual knowledge transfer and use is a very complex social interaction process (McAdam and McCreedy 1999; Nonaka, Toyama et al. 2000; Von-Krogh, Ichijo et al. 2000).

To Davenport and Prusak (2000) "knowledge is a fluid mix of framed experience, values, contextual information, and expert insight that provides a framework for evaluating and incorporating new experiences and information".

Other reference authors like Polanyi (1958) associate knowledge to action. He says that "knowledge is the ability to act". Nonaka and Takeuchi (1995) explain that knowledge is created by the flow of information associated with the beliefs and commitments of those who possess it.

In the view of Nonaka and Takeuchi (1995), knowledge is created within the company to make it more successful and to keep it on the market, competitive and ahead of its rivals.

Coulson Thomas (2002) remarks that today's organisations do not compete in terms of products, services or technology, but in terms of know-how, processes and values.

Related to knowledge another concept has emerged; Armstrong (2001) has studied the concept of intellectual capital as "the stocks and flows of knowledge available to an organisation". For this researcher, intellectual capital represents therefore a frame for three main elements:

The Human Capital, which represents the creative force of every organisation, in terms of skills, competencies and knowledge. The Social Capital, representing the relationships inside and outside the organisation which enhance the human capital potential. The Organisational Capital (i.e., materials, databases, manuals), which means what the organisation (and not the employees as individuals) actually possesses.

The human and the social capital still far from satisfactory in the most part of organisations. The existence of learning groups, the development of knowledge networks or communities of practice is still exceptions in few organisations.

According to Botha, Kourie and Snyman (2014, p. 9), "knowledge is considered a key part of the strategy to use expertise to create a sustainable competitive advantage in the present business environment." The main objective is to realize how using this knowledge can help to innovate and to adapt quickly to new circumstances, assuming that these issues have to be analyzed in the context of human and organizational dynamics.

It was also important to understand the nature of knowledge - tacit or explicit. Frappaolo and Wilson Todd (2000) refers that tacit knowledge is highly personalized, context sensitive and informal, and very hard to measure and manage. It includes know-how, intuition and informal communications that make up a large part of the organisation's culture. On the other hand, we have explicit knowledge (Nonaka 1994), as an object that can be codified and distributed outside of the individual who created it (Fahey and Prusak 1998).

Several authors (Clark and Staunton 1989; Blackler 1995) refer that implementing organisational innovation practices requires not only the translation of new knowledge from its abstract formulation into an organisational setting, it also requires its practical embedding in systematic routines and working practices and its "enculturation" in shared understandings, norms and values. As we shall see in the field research organisational routines are learning processes involving people doing things and solving problems, reflecting on what they are doing, and doing different things (or doing the same things differently) as a result of the reflection.

This perspective on routines is consistent, in several ways, with the work of Nonaka and Takeuchi on knowledge creation (1995). They claim that "seen from the vantage point of organisational knowledge creation, double-loop learning is not a special, difficult task but a daily activity for organisations". They also argue that change in organisations does not simply consist of responses to the external environment, but also consists of internally generated knowledge. Finally, they argue that there are four modes of knowledge, and that the interconnection of these four modes in a continuous spiral represents the process of knowledge creation.

Finnaly, the authors Phelps, Heidl and Wadhwa (2012) after an extensive theoretical and empirical review point out the importance of social relationships and networks in the processes of knowledge creation, dissemination and use.

METHODOLOGICAL STRATEGY

To accomplish the empirical work was applied the Action Research methodology and the main technique to collect data was group recall sessions with all levels of employees of the organization. Action research (AR) was the applied methodology for this study because we didn´t want to discovered generalizations, but contextual findings and rich insights and AR allows a deeper analysis and a different understanding of complex organizational problems (Sousa 2010, 2013).

Data was collected through interviews to top management and group recall techniques, which were also used to knowledge sharing among the researcher and the organizational actors.

The group recall technique can be framed in social research and the process is similar to focus group process. It gives the researcher the opportunity to hear detailed revelations about people's thoughts, ideas, and experiences. It has potential to illuminate workers' contrasting opinions and experiences and to help them getting to know better the organization, and sharing with colleagues their experiences and work practices.

The combined contributions of participants in the group recall might point up new directions and questions changing the trajectory of the session. There is then potential for the researcher to explore their remembrances allowing space for participants to articulate their own ideas.

The practical issues of group recall sessions are mainly the logistics of getting all participants together at the same time in the same room, prepared to talk on the same subject and the concerns that some contributions might become lost in the wider debate.

One main constrain about the use of this technique is the danger of censoring: individuals hold back the contributions they wish to make, conforming instead to an apparent consensus or to the opinions of a self-appointed "expert" within the group. Another concern is the possibility for participants to exaggerate their accounts in order to "impress" others. However, and because several participants could belong to the same work unit, this was unlikely to happen since the other members of the group would function as a control elements.

Collecting Data and Sharing Knowledge through Action Research

In this research, the group recall technique was used in order to share the knowledge among the researcher and the organisational actors. This is a qualitative technique created by Professor Peter Totterdill and his research team in Nottingham Trent University and applied in their researches.

The group recall technique can be framed in social research and the process is similar to focus group process. It gives the researcher the opportunity to hear detailed revelations about people's thoughts, ideas, and experiences.

It has potential to illuminate workers' contrasting opinions and experiences and to help them getting to know better the organisation, and sharing with colleagues their experiences and work practices.

The researcher conducts group recall sessions with a number of workers from one organisation at the same time in the same venue. It is a space for knowledge sharing through dialogue and it is capable of generating particularly high quality data.

Two main actors participate in the group recall process: the researcher, who has the capability to decide the design and direction of the research; and the researched, who can condition the success of the research with his involvement and participation in sharing their remembrances about their work experiences.

The group recall session is influenced by the number of participants, that should be no less than 3 and no more than 5 so that everyone has enough space to express their ideas and remember their work stories, practices and experiences.

The combined contributions of the participants of the group recall might point up new directions and questions changing the trajectory of the session. There is then potential for the researcher to explore their remembrances allowing space for participants to articulate their own ideas.

The practical issues of group recall sessions are mainly the logistics of getting all participants together at the same time in the same room, prepared to talk on the same subject and the concerns that some contributions might become lost in the wider debate.

One main concern about the use of this technique is the danger of censoring: individuals hold back the contributions they wish to make, conforming instead to an apparent consensus or to the opinions of a self-appointed "expert" within the group. Another concern is the possibility for participants to exaggerate their accounts in order to "impress" others. However, and because several participants could belong to the same work unit, this was unlikely to happen since the other members of the group would function as a control elements.

Moreover, its important to point out that, whatever research tool is used for data collection, the power over data analysis remains in the hands of the researcher.

Finally, and as an output from this research, it's important to identify the researcher's key responsibilities associated to group recall technique.

Table 1. Researcher's Group Recall Responsibilities

Researcher Role	Responsibility
Organizes the group recall sessions, focuses on communications between and among group recall participants. This is accomplished in informally face-to-face sessions.	Clarify communications Draw out the reticent Ensure that dissenting points of view are heard and understood Keep discussions on the topic Reconcile opposing points of view Ensure and articulate a valid purpose for the group recall realisation Stimulate interest in the group recall Organize face-to-face group recall sessions Stimulate organisational actors' participation Arrange for communications support Obtain official support when appropriate Communicate the contributions of the group recall sessions to the formal organisation

To assume a concept of knowledge share, I have analysed the existing literature and have concluded that it did not reveal knowledge sharing as a well-defined concept. Therefore, it is necessary to establish what we mean by knowledge sharing.

The term knowledge sharing implies the giving and receiving of information framed within a context by the knowledge of the source. What is received is the information framed by the knowledge of the recipient. Although based on the knowledge of the source, the knowledge received cannot be identical as the process of interpretation is subjective and is framed by our existing knowledge and our identity.

Knowledge sharing intrinsically implies the generation of knowledge. For instance, in face-to-face communication, an effective mechanism for gaining knowledge is to request help from another, i.e., someone who may possess the knowledge or expertise required. This request may lead to a conversation that will facilitate the creation of new knowledge. This suggests that in face-to-face interactions, conversations can be an effective conduit for knowledge sharing.

Conversation is the most common mean that facilitates the transfer and development of the more deeply rooted tacit knowledge. The context is built through communication and is enabled by a shared perspective, by language and a common understanding.

Sample

The empirical research involved a Portuguese company from the service sector, particularly engineering. A first meeting occurred in EFACEC (General Quarters), beginning the exploratory phase. In this meeting the innovation office manager explains the ongoing organisational innovation project.

The deep research began with the groups recall (3/4 elements per group): two with the technicians and one with the managers (head of departments). Also was made an interview with the General Manager of the organisation.

All the participants were selected by the Innovation Office Manager and the General Manager of EFACEC Robótica.

RESULTS

The company is a European leader for water heating manufacturing. Located in the North of Portugal and having about 1000 employees, the company has produced a major turnaround in financial and business performance during the late 1990s and early 2000s and continues to do so. It is a private company with a big focus on innovation.

Analysing the type of organisational innovation according to the OCDE (2002) definitions the studied company is working on:

First axis - restructuring production and efficiency processes, business re-engineering, flexible work arrangements, greater integration among functional lines, and decentralization.

Second axis - human resource management practices, flexible job design, employee involvement, and improving employees' skills.

Third axis - product/service quality-related practices, total quality management and improving coordination with customers/suppliers.

The company is implementing a system like TPS – Toyota Production System – with a holistic approach through the optimization of not only partial processes and departments, but all course and organisational units, especially their teamwork.

The Production System (PS) implies a systematic implementation of a multitude of devices designed to contribute to the improvement of quality, costs and delivery. The implementation of the Production System is best described through a phase model beginning with preparation, then stabilization and finishing with reduction. It is the basis for every PS-oriented project work whether in the reshaping of existing production lines, the planning of new lines or in the product creation process. The central idea of the PS is to develop and deliver the right part at the right time in the right amount and with the required quality.

To accomplish the principles of the Production System, the employees consistently demonstrated creativity and innovation – activities that placed an emphasis on the ability to engage in continuous learning and a deep knowledge of the organisation.

In this context to operationalize the knowledge sharing process the company has developed several mechanisms for creating and sharing knowledge - workshops, workgroups with workers from different sections and departments, suggestion boxes, and communication corners that are spaces for exchanging ideas, opinions and thoughts, but also to present structured knowledge through documentation, videos and other means.

The knowledge sharing process is explicit in the next figure, where it's possible to analyse the mechanisms and tools implemented to facilitate and potentiate the process.

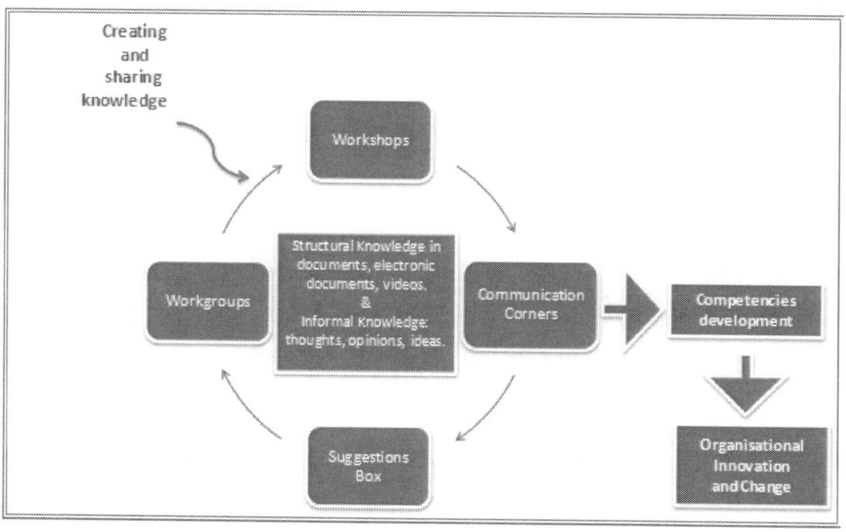

Figure 1. Creating and Sharing Knowledge.

Workshops are exceptional vehicles for bringing together employees from different areas to discuss an issue – i.e., a problem or the discussion of a new idea. These workshops can be scheduled on a regular basis during a period of time or they can be an isolated event set to share opinions and insights.

Sometimes it is necessary to invite outside speakers to these sessions. Agents who are in direct contact with customers, customers, specialists in some kind of area and other experts.

The invitation of customers for these workshops has an important role whenever the company is looking for new ideas or ways to improve products. It is an opportunity to share knowledge with people who do not know the technical aspects of the products, but can give important ideas about their functionalities, generating valuable shared insights.

Workgroups are created according to the needs of the organisation. Sometimes a well-defined problem statement is discussed, and the workgroup makes the necessary analysis and review, formulating recommendations for going forward.

At the moment, several workgroups of TPM (Total Production Management) where created, to improve the efficiency of the machines. These workgroups are very important to improve efficiency because Technicians undertake all the maintenance tasks, solving machine/equipment problems, while the Operators did not know how to assist them.

With TPM, company involves workers across the factory to ensure effective equipment operation in order to prevent breakdowns. TPM affects areas such as overall equipment effectiveness, total process reliability and total quality management and it's a process that requires the commitment from all the workers and Managers.

Workgroups are also established for strategy development and setting future directions. Cross-functional projects, training and program launches provide other ways to create workgroups and the share knowledge.

Another, but less standard, mechanism for sharing knowledge is communication corners. A simple 30-minute weekly meeting or a random meting when some kind of situation occurs can be invaluable. The idea is to jointly look at the operating results and discuss them, trying to understand them and finding new processes to reduce time or costs.

In these communication corners, the company presents operating data exhibits focusing on results and operating issues. This facilitates knowledge sharing across the company. All areas present their key operating results, creating a healthy challenge because the results are associated to an annual reward for the best team.

Competencies' development also helps to create a culture of knowledge sharing and the company has several training routines. The main goal is to create a potential - they invest in their workers so they can assume more responsibilities and become more knowledgeable.

Training assumed a big importance when the PS began to be implemented because everything was new: new principles, new practices, and new processes. This was the critical point of change. At that moment, the knowledge sharing performed a nuclear role, not only to disseminate knowledge about the system, but also to reduce resistance.

The next figure shows the main ways of learning and sharing knowledge by the employees. They share knowledge with their colleagues in several different and systematic ways creating opportunities to learn and maximize the organisation's ability to meet their needs and generate solutions and efficiency that provides the business a competitive advantage.

Embedding a culture of knowledge sharing and reuse is perhaps the most important challenge for Managers. It is less about managing knowledge and more about managing workers whose work depends on what they know and what can learn from others.

The process of applying Operator's knowledge with the help of the Technicians together with experimentation, observation and dialogue techniques, which allow the adaptation of existing knowledge to new and novel situations.

This represents an important and undervalued source of learning in the factory, and the processes of learning by observing are crucial for the new workers. They learn through socialization, observation and practice.

"Our instruction sheets of operating procedures and competencies tables represent a form of explicit knowledge in the plant, which can be used by the workers. But first they need to learn with the older workers or even the shift Managers how to use our work routines" (Group recall – Production Managers).

To share more objective knowledge like rules, procedures and routines, the company uses several techniques:

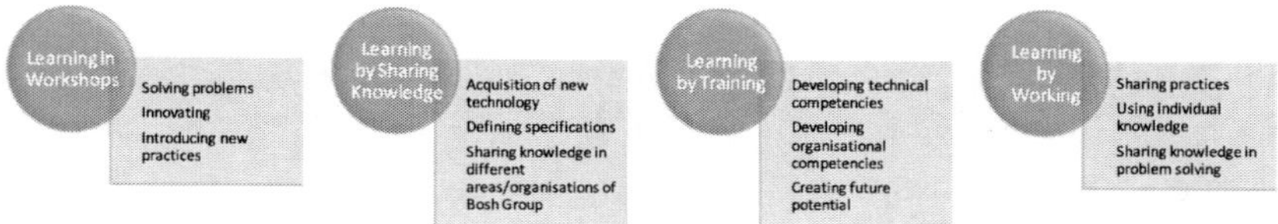

Figure 2. Learning Processes in a knowledge organization.

"We created a procedure sheet that new workers should follow" (Group recall – Production Managers).

Knowledge transference is often based on the organisation's explicit knowledge, in this case procedures sheets. However, they also have knowledge databases for quality problems, and solutions and others repositories where they store information and documents that can be reused and shared like, for example, product specifications, manuals, and other information regarding production.

In each section of the factory there is a computer where it is possible to search the databases, but even if it could be possible to use information technology for partial transference of explicit knowledge, without the tacit elements that underpin it, it will be impossible to develop a full understanding of what this knowledge means.

Contemporary literature is consensual about the fact that the sharing of tacit knowledge via information technology systems is extremely difficult, if not impossible, to achieve.

Explicit knowledge can be expressed in formal and systematic language, and shared in the form of data, specifications and manuals. Tacit knowledge is rooted in actions, procedures, routines and values. These characteristics make it easier to communicate explicit dimensions of knowledge than the tacit knowledge.

In many situations, tacit knowledge cannot be wholly converted into explicit. For instance, life and work experiences and all the knowledge those workers develop and store along the years.

It seems to be easier to share technical knowledge, because it is already explicit in manuals and it is easier to explain, then organisational knowledge that was accumulated along the years by the workers through work practices and routines.

"The technical individual knowledge is easier to transmit then managing principles. The procedures of some areas are already explicit. Nevertheless, it is not possible to transmit all the knowledge that we have. It is rooted in our experiences" (Group recall – Department Managers).

Department Managers focus that the company has that kind of problem when one worker is leaving within the organisation. They know that it is possible to share a percentage of all the knowledge rooted in workers. However, it is not possible to share all of it.

"In situations where worker are leaving the company, we defined a period of time for them to work overlapped to transmit the knowledge to another colleague" (Group recall – Department Managers).

But how can we make it explicit? Davenport and Prusak (1998) stated that the transfer of knowledge can be made by formalized transfer mechanisms and informal exchanges.

The formalized transfer methods include documents, databases, Intranets and GroupWare. Informal exchanges refer to the more casual events that usually take place face to face such as a conversation.

Ways of making the knowledge explicit:

Written: through e-mails, documents and discussion groups.

"All the procedures are available through documentation and in the intranet. Workers can access computers in each section to consult the information or ask the section's Manager to access the information for him, because some Operators do not know how to access the intranet and make the search or even how to use the computer" (Group recall – Technicians).

Visual: using models, illustrations or data visualization tools.

"The information is all registered in photographs and displayed in the sections' placards. The same happens with instructions, work plans, maps and tables, so that they are easier to read and understand" (Group recall – Technicians).

They take photographs when some changes in the plant occur and then they display it in the plant showing how it was before and how it is now. This very powerful technique helps to involve the workers in the organisation and in their work. They create emotional liaisons to their workstations, when they analyse all the changes that they have faced and overcome.

Production Managers also referred other kind of information displayed:

"In the communication corners we place all the important information: efficiency levels, competencies matrix, instructions and productivity data" (Group recall – Production Managers).

Spoken word: through voice mail, recordings, the telephone or person-to-person interaction.

"Communication corners are used for meetings. In some sections, the meetings are held weekly, with the goal to analyse all the issues that occurred in the previous week. With this we look forward to eliminate "Mr. Rumour" and involve the workers in all the factory situations and problems" (Group recall – Production Managers).

Video/ observation: video databases, body language, master-apprentice relationship, video conferencing.

"For instance, we have problems and solutions databases and quality databases that are accessible to all workers, in each section of the plant" (Group recall – Production Managers).

Combination: technologies adopted that include some or all of the previous.

"When we have problems we register them in an internal tool together with all the information related to the problems" (Group recall – Technicians).

One main reflection from group recalls is that individual knowledge, if not shared with others, will have very little or none effect on the organisation. Therefore, one of the important tasks for organisations is to facilitate the process of interaction between employees promoting and encouraging the use/share as well as using the knowledge gained and stored in the form of explicit knowledge.

A fact with major interest is that the knowledge sharing routines involve not only internal actors, but also external ones like, for instance, customers. They bring fresh and new ideas especially to improve products, as the Department Managers point out:

"When we need to improve our products, we organize workshops to discuss the new possibilities and functionalities that they could/should have. We invite several people outside the organisation, mainly costumers. Sometimes they have very simple ideas that we would never have remembered because we were focus on a more complex problem. Their ideas give us a new and fresher perspective of the possibilities that our products could have and this is phenomenal because our costumers normally identify themselves with the new improvements that emerge from this sharing" (Group recall – Department Managers).

Even a specialist in a certain area can help the discussion and create some knowledge that can help to implement a new practice, tool or technology:

"The workshops sometimes have an external moderator, someone with specific knowledge. This helps us to develop ourselves in an organisational and technological way. Sometimes it is a specialist belonging to the group (from another factory with a different line of products), other times it is a consultant or someone from the Academy" (Group recall – Department Managers).

When they have a production problem, the workshops involve only internal actors from different sections of the plant so that together they can all find a solution or a way to minimise the consequences of a problem.

"When we have some kind of problem, the moderator is internal and the idea is to involve all workers related to the situation under discussion. This workshop has very well defined goals and if the workshop finishes without having solved a certain problem, a schedule is defined with actions delimited in time until the problem or the situation is solved.

The greatest advantage of workshops is that people who participate in them are an integrant part of the problem or situation. An example of a problem that we discussed in a workshop was regarding auto-quality – we intend to reduce the incidents of line 1" (Group recall – Department Managers).

The workshops can be seen as knowledge creation processes, such as communities of practice or other processes of linking workers to others with expertise. Relational competences are a key to the capture, use and creation of knowledge and learning within organisations.

However, it is important to think about the key challenges involved in building strong relationships between employees. The management challenge is to create an environment that values sharing knowledge and the personal challenge is to be open to the ideas of others, and also be willing to share ideas.

Conclusion

The company began the whole innovation process by implementing the Production System, a very structured system with several tools adapted to all organisational dimensions.

One of the critical factors of success is the top management involvement in all the processes, and the willingness to create and implement a culture of innovation and change. This culture is being created on a daily basis, creating habits and behaviours of participation, communication and involvement in all aspects – this constant change involves both micro and macro changes.

During the group recall sessions, almost all the actors have made some kind of suggestion of change, not only involving their workstations, but also the organisation itself. This culture of innovation and participation is deeply integrated in the organisational life.

When we analyse the routines for creating and sharing knowledge, we can find several mechanisms used to facilitate the share: suggestion boxes, openness to make suggestions to the Managers, several types of workshops where employees from different sections participate, and several transversal projects of improvement, quality and maintenance.

With great visibility, the workers use the suggestion boxes as a space where they can uncover new ideas that help improve the organisation.

Cross-functional workshops and meetings are a crucial space to share perspectives and to make discussions that provide invaluable knowledge.

Organisational actors share their opinions and insights, as well as their own questions, sharing and creating new knowledge.

For added impact, outside specialists and even costumers participate in these sessions. Their perspectives can be refreshing and break down the thinking routines of internal workers.

Transversal projects like TPM or projects related to quality systems also help to develop workers' competencies with the share of their knowledge and experiences.

Like Reid argued in his research that "the most effective way to disseminate knowledge and best practice is through systematic transfer" (2003). The company has been creating a culture of knowledge sharing by implementing these routines and promoting collaboration in a systematic transfer.

To show its commitment for sharing knowledge, the company created a reward system taking into account workers' contribution and their participation in the organisational life. The reward system was created to benefit workers who contributed to and used a shared knowledge base. Their contribution is also recognized through information displayed in the plant.

Finally, it's also important to stress the importance of the sharing during the training programs that have prepared Managers and workers to work within the new set of organisational dynamics imposed by the Production System.

FUTURE RESEARCH DIRECTIONS

New expectations of knowledge sharing and use are emerging as this research field is becoming strategically important for business organizations. In this context, a consistent framework needs to be developed.

Directly related to the findings from this research, other aspects could be developed by the company:

The model created for facilitating the creation of a knowledge sharing culture has not been implemented. An implementation of the model and its implications on the organization is an interesting topic for future research.

The creation and applicability of an instrument to diagnose the knowledge profiles also seems to hold potential. The results could be used in the training plans in order to develop organizational actors' potential for managing knowledge.

Further research could also be undertaken:

Studies on knowledge integration across organizational functions and in other types of organizations.

Studies that develop and test a theoretical framework that relates knowledge integration mechanisms, situational characteristics and organizational outcomes.

Studies that analyses the capabilities of employees' informal networks in order to achieve efficient integration of knowledge into work practices.

Furthermore, future studies are required to determine the importance of different types of knowledge sharing for different organizational activities. Finally, more research is needed on the facilitation of the different types of knowledge sharing.

REFERENCES

Armstrong, M. A. (2001). Handbook of Human Resource Management Practice, 8th edition, Kogan.

Blackler, F., 1995. Knowledge, knowledge work and organisations: an overview and interpretation. *Organisation Studies*, 16(6), pp.1021–46.

Botha, A., Kourie, D., and Snyman, R. (2014). Coping with continuous change in the business environment: KMand KMtechnology. Elsevier.

Clark, P. and Staunton, N., 1989. Innovation in technology and organisation. London and New York: Routledge.

Coulson-Thomas, C., (2002). Transforming the company. 2nd ed. London: Kogan Page.

Davenport, H. T. and Prusak, L., (2000). Working knowledge: how Organisations manage what they know. Boston: Harvard Business School Press.

Fahey, L. and Prusak, L. (1998). The eleven deadliest sins of knowledge management. *California Management Review*, 40(3), 265-276.

Frappaolo, C. (2008). Implicit knowledge, *Knowledge Management Research and Practice,* 6(1): 23-25.

McAdam, R. and McCreedy, S., (1999). The process of knowledge management within organisations: a critical assessment of both theory and practice. *Knowledge and Process Management*, 6(2), pp. 101-112.

Nonaka, I. and Takeuchi, H., (1995). The knowledge-creating company: how Japanese companies create the dynamics of innovation. New York, NY: Oxford University Press.

Nonaka, I. (1994). A dynamic theory of organisational knowledge creation. *Organisation Science*, 5(1), 14-37.

Nonaka, I., R. Toyama, et al. (2000). SECI, Ba and Leadership: a Unified Model of Dynamic Knowledge Creation. *Long Range Planning*, 33: 5-34.

Phelps, C., Heidl, R., and Wadhwa, A. (2012). Knowledge, networks, and knowledge networks a review and research agenda. *Journal of Management,* 38(4), 1115-1166.

Polanyi, M., (1958). Personal Knowledge. Towards a Post Critical Philosophy. London: Routledge. 428 pages.

Sousa, M. J. (2013). Knowledge profiles boosting innovation. *Knowledge Management,* 12 (4), pp. 35-46.

Sousa, M. J., (2010). Dynamic knowledge: an action research project. The *International Journal of Knowledge, Culture and Change Management,* 10(1), pp. 317-331.

Von-Krogh, G., K. Ichijo, et al. (2000). Enabling Knowledge Creation: How to Unlock the Mystery of Tacit Knowledge and Release the Power of Innovation. Oxford, Oxford University Press.

In: Knowledge Creation and Transfer
Editor: Edwin Matos

ISBN: 978-1-63485-656-0

Chapter 3

CO-DEVELOPING A KNOWLEDGE EXCHANGE NETWORK TO FACILITATE ACCESS TO CLINICAL BEST PRACTICES DURING CARE TRANSITIONS OF PATIENTS WITH CHRONIC DISEASES: AN EXAMPLE IN ONCOLOGY

Hélène Lefebvre**, *Isabelle Brault*, *Marie-Josée Levert*, *Odette Roy*, *Sophie Alarie*, *Michelle Proulx* and *Maryse Larivière

Faculty of Nursing, Université de Montréal, Montréal, Quebec, Canada

ABSTRACT

Despite numerous Quebec initiatives to improve transitional care for patients with chronic diseases such as cancer, clinical teams and patients still encounter some difficulties during these transitions. This article presents a review of the literature on the involvement of clinical teams and patient-as-partners in the changes made to clinical practices. It also highlights the main issues that arise during the implementation of a

* Corresponding author: Hélène Lefebvre, PhD, Professor, Faculty of Nursing Université de Montréal, Montréal, Canada, CP. 6128 Succ. Centre-Ville Montréal, Quebec Canada, H3T3J7, helene.lefebvre@umontreal.ca.

knowledge exchange strategy with a *Forum for Knowledge Exchange* (FKE) that aims to guide clinical teams and patients to improve the continuum of care in oncology. The FKE is an interactive space for researchers, nurses, other healthcare professionals, and patient-as-partners where they can share their knowledge, both tacit and scientific. This study uses a participative methodology and is conducted in six clinical settings that span the continuum of care in oncology. Patients-as-partners are an integral part of the implementation process, as active participants rather than in a simply advisory role. This strategy should help improve transitions between the different phases of the care continuum. The co-construction of knowledge stemming from this clinical research study should lead to a renewal of clinical practices and thus guide clinical decision-making in multidisciplinary teams and patients.

Keywords: nursing, oncology, information and communications technology (ICT), care transitions, participatory research, patients-as-partners, patient participation

INTRODUCTION

Numerous Quebec initiatives have sought to improve transitional care for patients with chronic diseases such as cancer, but they have not yet led to the expected outcomes [1]. Clinical teams and patients need to increase their involvement at different stages of the care process in order to optimize the experiences of patients. Managing care transitions for oncology patients is at the heart of the issue for nurses, other health care professionals, and patients in order to improve the quality of care they receive. The planning of care in the continuum, follow-up, support, and communication between professionals are all still poorly coordinated and present significant shortcomings [2], especially since in Canada, there are no rules governing care transitions such as the planning of the discharge during the transition from one step to another. According to the scientific literature, good management of discharge planning would have positive effects on the quality of care and the physical and emotional well-being of patients and their caregivers. A detailed hospital discharge plan would reduce the length of hospital stays [3], diminish readmission costs [4], and optimize the transitions between different levels of care [5]. According to Naylor and collaborators [6], a 10% decrease in the rate of re-hospitalization was observed in patients benefitting from discharge planning, which included follow-ups by nurses on the management of physical symptoms, the taking of medication, sleep assessment, pain relief, and the emotional state of patients and their caregivers. Some authors underline the key role played by educating patients/caregivers, coordinating services, and implementing regular follow-ups as essential elements to optimize hospital discharge planning [7, 8].

This article presents a review of the literature on the involvement of clinical teams and patients-as-partners in the changes made to clinical practices through a community of practice. It also highlights the main issues that arise during the implementation of a knowledge exchange strategy with a *Forum for Knowledge Exchange* (FKE) that aims to guide clinical teams and patients to improve the continuum of care in oncology.

A Community of Practice: Informed Decisions at Your Fingertips

Inspired by past practices such as mentoring, a community of practice is an informal network of persons who share their passion and expertise in a common professional field [9]. However, more formal and structured communities of practice have developed, backed by organizations (St-Onge and Wallace, 2003). According to Prax [10], the community of practice is based on the flow of knowledge and constitutes “an ideal forum for collective intelligence, innovation, creation and collective knowledge.”

> “A community of practice is a “space” for co-development where members can exchange knowledge in order to improve their practice.” [11]

According to Wenger, the community of practice is essential for organizations in an increasingly knowledge-based economic reality. For Davel and Tremblay [12], communities of practice are a strategic path to developing and sharing tacit knowledge and are vital tools in the knowledge economy. They are viewed as informal groups defined not only by their members but also by their methods and interpretation of events. Members of this type of community maintain relationships with one another, participate in activities over time, and develop relationships with other communities of practice [13]. Knowledge transfer and the ways in which practices are perpetuated are the cornerstone of a community of practice. Its identity is dependent on the relationships and activities born of these social interactions, of specific skills and experiences in a dynamic process over time [13].

More specifically, the community of practice is a group of persons who collaborate (through interactive platforms such as forums, videoconferencing, email, etc.) in order to develop innovations or solve problems encountered in their specific professional practice, gradually leading to joint learning through the sharing of individual knowledge and expertise [14].

There Is a Community of Practice if….

A community of practice does not refer to a service or a network [15], rather it is a group of professionals who share knowledge, establish common practices, and develop them in a common field of interest [12]. Members freely exchange creative experiences and knowledge that promote the emergence of new approaches to problem resolution [16]. A community of practice is established when a group of people, informally linked and who share a professional practice or field of expertise and a common professional goal [17], exchange, share, and learn with and from each other.

There are three types of communities of practice: 1) Loosely structured and poorly organized informal communities; 2) more developed, established communities benefitting from organizational support and characterized by the development of new knowledge/skills in the members of their practice; and 3) structured communities of practice with strategic imperatives whose goal is organizational performance. The *Réseau infirmier, un partenaire de soins* (RIUPS) (The nursing network: a partner in care) falls into the second category.

A community of practice is defined as [9, 12, 16, 17, 18].

Table 1. Characteristics of a community of practice

A common interest
Established, mutual relationships that can be harmonious or contentious
A mutual commitment to doing things together
Knowledge of what others know, what they can do, and an eventual contribution to collective action
Shared methods of communication and a history
Shared ideas that influence members' vision of the world
Development of knowledge and expertise in a field that is improved by the quality of the interactions and common interests

THE FOUNDATIONS OF A COMMUNITY OF PRACTICE

Communities of practice are built on three principles [17]: mutual commitment; a common undertaking; and a shared inventory of resources that allows members to communicate and solve problems. The communities contribute to creating and transforming practices, ideas, and the organizational culture [19]. Mutual commitment is a fundamental factor to development [17] and determines its evolution, results, and longevity. Belonging to a community of practice is the result of a commitment to actions whose meaning is dependent on the relationship between the members of the community. Mutual commitment is founded on the complementarity of skills and the capacity of individuals to effectively combine their knowledge with that of others. The necessity of linking skill sets is particularly evident in communities where mutual commitment implies complementary contributions. Thus, learning occurs through an intellectual process on the one hand and through membership to the group and active participation in the interactions that occur on the other [9], evolving over time. These groups are normalized while remaining heterogeneous, each member's diversity contributing to a shared history [9], an understanding, and a coordination of a joint venture towards a shared objective. This process unfolds through the development of shared catalogues of ideas and modes of operation, of the shared product that is the result of the interactions between members.

For Wenger, McDermott, and Snyder [9], commitment is a central issue, whereas for Gherardi et al. [13], the methods of organization support the exchange and perpetuation of practices [13], placing the emphasis on the notion of productive and efficient practices that are not necessarily exempt from conflict. According to Gherardi and Nicolini [20], individuals must learn to obey the rules of a given practice. Socialization within a community of practice refers to the process by which novices adopt the language, the ethics and the culture of the community that supports their practice. Knowledge is integrated and circulated within the community, and, since learning is evidence of membership, it requires commitment and relies on socializations processes [12]. The process necessitates a mutual effort by novices and the community [15] as their command of the practices improves [13, 15].

However, certain favourable conditions to the creation and optimization of these communities of practice must be present in order to promote their development. These are:

1. Organizational support and managers [21].
2. Effective project management [21].
3. Support for training [21].
4. Efficient organization of the community [22].

Sometimes, practices can extend outside their communities to latch on to others by creating different networks of interactions within a group of communities of practice linked by interconnected fields [12, 20].

The development of a community of practice follows five stages (Figure 1) and is dependent on the environment, history, and cultural, economic, and sociopolitical contexts that can influence its development as well as the financial, material and human resources at its disposal, such as those related to organization [9].

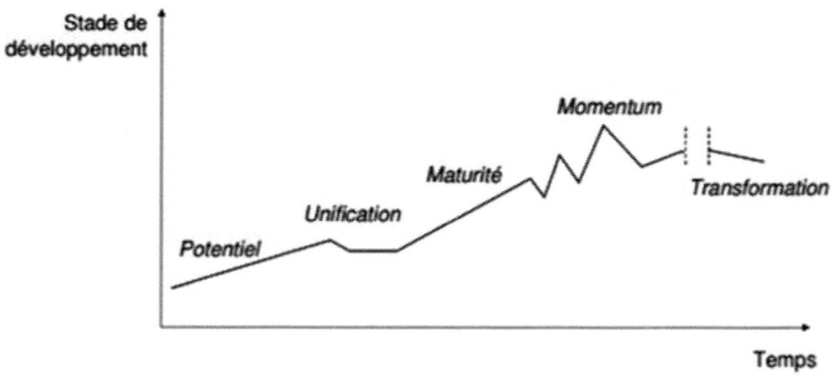

Source: Source: Davel and Tremblay (2011) adapted from Wenger et al. (2002, p. 69).

Figure 1. Phases of development of a community of practice.

During the first stage, Potential, prospective participants to the community of practice see the value in working together. In the second stage, Coalescing, the community is launched, spaces are defined, events take shape, leaders are enlisted, relationships are developed between participants, and the activities, which will eventually evolve, are determined [9]. As of this stage, documenting the process and activities is essential, as is coordinating the community, which requires time in order to identify the interests "of" and "for" the organization and to develop the technical knowledge required to moderate an online community and networking skills. During the third stage in the development process, Maturing, the community broadens the field of interest and evolves, often in a non-linear manner. Its values and objectives crystallize, as do its direction, roles, the limits of its scope of activity, and its relationships with other organizations/communities. Gradually, these steps are better defined and it is at this stage that the services of a documentalist are required. The Stewardship stage follows. Organizational tensions appear and it is imperative to maintain intellectual and emotional interest and stay abreast of the main developments in the fields of interest. Leadership and membership to the community can be realigned, and relationships and resources from outside the organization are often developed. The final stage of development, Transformation, is crucial because identity issues are rarely completely resolved and the development of the area of interest is not always unanimous. According to Wenger et al. [9], the best communities of practice are those that promote debates that lead to their effectiveness and productivity and spur major transformations, whereas others are gradually dissolved.

According to the scientific literature, participation in a community of practice leads to improved organizational performance [12], the development of a culture of sharing [12], specialized vocabulary [16, 23], and a framework [16, 23].

However, Snyder and Wenger [16] identified additional advantages of communities of practice such as contributing to a strategy, solving problems, sharing innovative practices, creating new products and services, developing professional skills, and recruiting new members.

In this sense, communities of practice tend to be, in certain fields, "more advanced" than the traditional teaching methods available and present an opportunity for quicker adaptation to the needs and realities of organizations, all while providing a way to distribute information that sometimes requires less logistical planning.

Learning Methods

Communities of practice arouse interest because they provide less conventional opportunities for learning. The European Commission [24] identifies three types of learning methods:

1. **Formal learning.** Traditionally takes place in an educational or training institution, structured (in terms of objectives, time or resources), is intentional on the part of the learner and leads to certification.

2. **Non-formal learning**. Structured learning (in terms of objectives, time or resources), taking place outside of an educational/training institution, non-formal, but is intentional on the part of the learner.
3. **Informal learning.** Learning that is not structured (in terms of objectives, time or resources), but results from daily activities and situations related to work, family life or leisure that can be intentional, but generally is not. This type of learning is generally random and unplanned.

Yet, communities of practice provide opportunities for non-formal and informal learning, depending on the joint venture linking them. Since, according to Jalali [25], learning methods and modes of communication are changing, and adapting to these new methods of communication is not a choice, but rather a necessity, communities of practice made up of a group of individuals who collaborate (for example through interactive platforms such as forums, videoconferencing, email, etc.) [14] are a promising way forward. In fact, given the context of budget cuts where time and travel are becoming a luxury organizations can no longer afford, and the fact that modes of communication are rapidly evolving, the community of practice presents a viable solution for the future, adapted to the realities faced by the new generation providing healthcare services to patients with chronic diseases, for example.

A Virtual Community of Practice

In light of the above, virtual communities of practice, specifically those that use the Internet as a means of communicating and collaborating in order to engage in a process of co-learning, are modern learning processes that aim to consolidate and document formal and informal knowledge, through a network.

> "Individuals learn in their work settings, where knowledge is acquired and shared in a less formal environment than that of a traditional training session. Interacting, exchanging information, discussing actual problems encountered during their practice with colleagues, exploring various ways to innovate, all create a situation where they are developing their knowledge." [26]

The Internet provides a loosely structured, effective means of communication that preserves records of the exchanges.

> "When a virtual community of practice is established, the online forum provides a place where participants can communicate and collaborate. Participants see the importance of exchanging information, of cooperating or collaborating to tackle a complex task, of solving a difficult problem that is nearly impossible to solve alone, or of formulating critical thoughts that require logical and viable answers." [26]

INTERACTION: THE FOUNDATION OF LEARNING

In a traditional community of practice, interactions are conducted in person, whereas in a networked community of practice, collaboration and interaction take place online, in conjunction with and as a complement to face-to-face meetings. In a virtual community of practice, face-to-face meetings gradually diminish and are eventually almost completely eliminated. Knowledge sharing can occur either synchronously or asynchronously. Synchronous communication occurs in real time (during face-to-face activities, videoconferencing, online chat sessions, etc.). Asynchronous communication takes place in delayed time and electronic forums are now used for this purpose. Participants no longer need to be present on the forum at the same time, and with Internet access, they can be in different geographical locations and still be able to browse the community of practice's electronic forum.

The Internet facilitates interactions by removing the constraints of time and place, and also creates a record of message history. The electronic forum provides a place for participants to communicate and work. They exchange information, cooperate, or collaborate [27, 28] to tackle a complex task, solve a problem, or stimulate reflection. In the healthcare field, medical knowledge evolves rapidly, and nurses, for example, have increasingly less time and resources for training. Thus, communities of practice are becoming indispensable to ensure their knowledge and skills stay up to date. Knowledge transfer occurs quickly and reaches a large number of people, who in turn contribute, through the community, to broadening the knowledge of others by participating and sharing their experiences.

Towards a Wider Community: The Involvement of Patients-as-Partners

Recently, communities of practice in the healthcare sector have benefitted from the contributions of a new partner: patients. Indeed, the relationship between patients and healthcare professionals has been one of the latter's concerns for many years [29, 30]. For a long time, this relationship was observed almost exclusively from the perspective of healthcare professionals and organizations. Consequently, access to patients' perspectives on the system and healthcare services was almost non-existent. Few studies have focused on the patients' viewpoints, despite results showing significant disparities between the perceptions of healthcare professionals and patients regarding needs and the quality of care [31]. In recent years, a study of the literature seems to indicate that healthcare professionals and researchers are increasingly interested in patients' perspectives and experiential knowledge of the care process [32, 33]. Since 50% of the population of Western countries is living with at least one chronic disease, and 80% of Internet users search for health-related information online [34], it is imperative to ensure they participate in the care process, and, consequently, in the communities of practice in order to adequately meet their actual needs. It is with this in mind that the *Direction collaboration et partenariat patient* (Patient collaboration and partnership department) of the Université de Montréal's Faculty of Medicine is developing a new relational model based on the partnership between patients and healthcare professionals: the *Montreal Model* [35]. It recognizes the "patients' knowledge, gained from living with health or psychosocial problems, their experiences, and their understanding of the care process, as well as the impact of these problems on their personal lives and that of their caregivers", which is complementary to scientific knowledge [36].

This relational model views patients-as-partners in all the decisions concerning them and considers them experts in the organization of care.

The Réseau Infirmier, un Partenaire de Soins (RIUPS) (The Nursing Network: A Partner in Care): An Evolving Community of Practice

In the current context of Quebec healthcare network reform, which makes access to training more difficult for nurses and other healthcare professionals, one of the objectives of the RIUPS is to increase the effectiveness of communities of practice. The clinical settings associated with the project have observed a significant interest in the RIUPS community of practice from those who wish to keep up with current knowledge and benefit from the expertise and tools from every care setting.

The RIUPS provides an opportunity for inter-organizational networking, access to specialized literature, and a forum for discussion that can be recommended to patients. The RIUPS is a place to exchange experiences, know-how, and scientific knowledge between patients, their caregivers and healthcare professionals. It focuses on what patients and their caregivers experience, feel, try out, witness, and learn throughout their care process. These experiences take place within the framework of the various relationships they develop with the many healthcare professionals they encounter over time and during their treatments. The RIUPS is not a reference website for professionals, but rather a forum for interactions to better understand the experiences of others. The RIUPS recognizes that the voice of patients and their caregivers is essential, needs to be heard, and should be taken into account by healthcare professionals, so they can adapt their clinical practice to the actual needs of patients and their caregivers.

It is in view of this that, since 2013, the research conducted by the Université de Montréal's RIUPS has focused on the best ways to make study findings and innovations developed by partner institutions available to nurses, healthcare professionals and patients. The strengthening of the collaboration between Université de Montréal's Faculty of Nursing and the institutions part of the healthcare network has enabled the development of an interactive nursing and interprofessional community that stays current by organizing knowledge exchange activities between nurses and other professionals, researchers, managers, decision-makers, and patients/caregivers[1].

Indeed, patient involvement is a promising avenue towards improving the health of the population and finding solutions to the problems the healthcare system is facing [37, 38, 39, 40, 41]. Experiential knowledge could contribute to improving the quality of care and of life for both patients and healthcare professionals.

[1] This study received funding from the Canadian Institutes of Health Research (CIHR-Fall 2013) through its Partnerships for Health System Improvement (PHSI) program and from the Health Research Foundation (HRF). Financial support was also provided by the Vice-President of Research at Université de Montréal.

> "Successful patient involvement, the level of patient participation at different stratum of the healthcare system, and the satisfaction expressed by all the actors who experienced this partnership approach are encouraging factors when considering long-term improvement of practices and of the health condition of populations with chronic diseases." [35].

The RIUPS was created with this in mind. The RIUPS is a community, which is supported, developed, and that benefits from organizational support characterized by the acquisition of new knowledge/skills by members in their practice. It falls into the second category of community of practice presented earlier in this chapter. The network is a clinical innovation, which includes an interactive portal, the Forum for Knowledge Exchange (FKE). The Forum is an interdisciplinary online portal that aims to bring together researchers, nurses, doctors, other professionals, managers, decision-makers, and patients-as-partners in a new space to share scientific and experiential knowledge with the aim of improving care. This tool provides support to healthcare teams to promote knowledge transfer and quick access to the best clinical practices in hospital discharge planning and care transition in general, and more specifically in oncology. The Forum for Knowledge Exchange, led by teams of professionals and patients-as-partners in clinical settings, aims to meet the needs to interact, exchange and share of actors in a same clinical field in order to co-construct new knowledge that is better adapted to the needs of people. As opposed to traditional platforms, which are more static and used mainly for filing documents, the Forum uses Web 2.0 technologies to promote interactions. Users have access, through a private link, to a documentation section and a collaborative work section (journal club, webinars, message boards, etc.). To optimize its visibility, the RIUPS is also on Twitter (https://twitter.com/RIUPS_FSI) and Facebook (https:// www.facebook.com/ riups.fsi).

The *Direction générale de cancérologie* (DGC), the organization in charge of implementing an integrated oncology network in Quebec (Canada), has focused its efforts on the steps preceding the cancer diagnosis and surgery to consolidate the continuum of care and meet ministry standards [42]. However, patients' reinsertion into the community has not yet been studied. According to the province of Ontario's Excellent Care for All strategy, "Reducing avoidable readmissions of patients discharged from hospital is an important area for improving the quality and safety of health care and making more effective use of health care resources in Ontario." p.6 [43]. This strategy highlights the importance of using evidence-based data to ensure better care transitions and avoid hospital readmissions [43].

Even though some studies have shown that using scientific knowledge has positive effects on patients' conditions [44], others have underlined the lack of impact on their health due to the gaps in sharing and implementing clinical practice guidelines [45]. These findings have led to the development of knowledge transfer strategies to facilitate their integration into the renewal of clinical practices. Among these strategies, key elements that encourage knowledge exchange are developing an organizational culture that fosters knowledge exchange [46] and identifying individuals in leadership positions as "champions" [7, 47, 48]. However, the exchange of new scientific knowledge in clinical practice remains underused by healthcare professionals in general [49, 50] as well as by nurses [50].

The Forum for Knowledge Exchange, hosted by nurse leaders and clinical knowledge development teams from clinical settings in a context of discharge planning and oncology care transitions, provides a way to group knowledge so all participants can access it and work together to develop new knowledge.

Hospital discharge planning refers to the various methods implemented in institutions to prepare transitions in patient care (e.g., identifying needs, assessing capabilities, recommending the optimal service). The transitions reference the different steps cancer patients experience in various oncology care contexts, including, among others, when they are discharged, relocated within the same institution, or reintegrated in the community [16]. This project takes into account the perspective of the patients and their caregivers from the beginning of the Forum for Knowledge Exchange and throughout its development. Patients and their caregivers are encouraged to participate in the activities provided on the Forum and to reflect with healthcare professionals on the best ways to improve hospital discharge planning and transitions in oncology care. Their mandate will eventually be broadened.

The Forum for Knowledge Exchange is established and available online (riups.org). Contributions are made on a weekly basis and interest is sustained. The administrative infrastructure is now stable. Contact persons are identified for each setting to ensure the steady transfer of information to the teams and improve the flow of information between the RIUPS and the members of the designated teams. The recent involvement of members of the designated teams in the activities of the RIUPS outside their settings indicates that the community is gaining momentum and is thus on the road to "Stewardship". Tools have already been added to the Forum by the clinical teams, and using the Forum for inter-team collaborations to develop a transition planning tool is in progress. Figure 2 present the governance structure of the RIUPS.

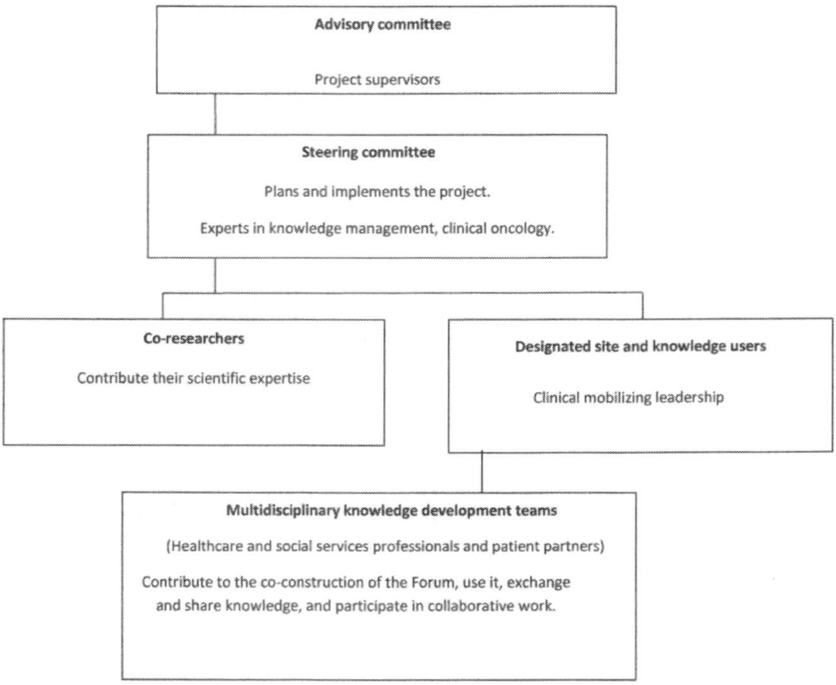

Figure 2. Project governance structure.

Five patients-as-partners are continually involved in the community of practice and have become a driving force in the co-construction of the clinical project. They participate in the steering committee, in meetings of knowledge users (nurses who host the clinical teams, including the patients-as-partners from their clinical setting), in patients-as-partners committees, etc. as stakeholders in the project. They have created videos and written articles available on the Forum, they contribute to the message boards, and play an important leadership role in the development of upcoming activities and the promotion of the project in the settings. They are ambassadors for the RIUPS within clinical settings and the community. They recruit new patients-as-partners to take into account a range of situations. Initially, patients-as-partners were referred by the clinical teams, but new patients-as-partners have joined the project through word of mouth. The involvement of patients-as-partners plays an essential role in the path to permanence of the RIUPS community of practice. Moreover, their participation has evolved and they have developed a leadership during the implementation process, which has a positive impact not only on the development of the community of practice of which they are members in equal standing to healthcare professionals but also on the clinical settings that have already begun to rethink their practice. Indeed, the role of the patients-as-partners and their approach during meetings is leading other teams to a new level of participation in the project, within an important context of organizational restructuring. The willingness of professionals to consider the perspective of patients during their care is seen as value added and an opportunity to improve care and services, both for patients and healthcare professionals. Collaborative work between clinical settings and patients occurs within the community of practice and the role of the research team becomes secondary.

Recent observations on the activities of the RIUPS show that meetings between patients-as-partners and settings promote exchanges in a process of mutual reinforcement. Relationships between patients-as-partners and various clinical settings are forged increasingly easily. The patients-as-partners' leadership in the community of practice is contributing to the optimal development of the RIUPS. Working in partnership with patients within a relational process contributes to knowledge co-construction, bilateral exchanges, shared power, equal relationships, teamwork, co-learning, autonomy, openness, empathy, mutual trust, satisfaction for patients and professionals, and improved communication. It provides a broader vision based on mutual objectives, centered on needs and expectations, individual or personal, that is rooted in reality. It is apparent that the RIUPS community of practice has positive effects on the quality of care in terms of effectiveness, through a better assessment of needs, increased efficiency, and improved coordination of care, all while facilitating the adaptation and improvement of the quality of care.

However, this community of practice also presents some challenges. Partnership implies a shift in paradigm at the clinical and organizational levels. Indeed, one must get to know the person in front of them; develop the capacity for adaptation and an openness to change; and unite the multiple competing visions, values and beliefs. Moreover, working in partnership within the community of practice requires a redefinition of everyday priorities according to the complexity, number, and lack of resources in the healthcare system, and a flow of information between professionals and patients (coordination, continuity of care).

CONCLUSION

The development of a community of practice, of which patients-as-partners are an integral part, contributes to the improvement of the quality of care and services provided to patients and their caregivers during hospital discharge planning and transitions in oncology, through co-construction and a more widespread use of scientific knowledge and the renewal of clinical and organizational practices. The various actors who are part of the project are engaged in an ongoing process of co-construction and knowledge exchange, which aims to transform their clinical practices and have a positive impact on patients.

Policy-makers participating in the project will be motivated to develop advanced healthcare policies by relying on the project's findings and to influence the decisions made in their institutions, thus contributing to the improvement of decisions made by professionals as well as by patients themselves. The use of the Forum will have the effect of locally transforming clinical practices in discharge and transition planning in oncology and placing them within a global, inter-institution continuum of care. Eventually, the use of the Forum should be expanded to include other health issues in order to better meet the actual needs of patients.

Virtual communities of practice will probably become increasingly prevalent in modern organizations given the financial and organizational contexts, which are making replacements and travel more difficult, and referring to best practices is becoming increasingly important. Since emerging healthcare professionals are already immersed in the social media world and the shift is now well underway in education, the health sector will have no choice but to follow suit in order to continue to provide services of the highest quality and to adapt to new realities. In this context, virtual communities of practice present a quick, inexpensive path to best practices and to a network enabling the instantaneous transfer of knowledge between healthcare professionals. Even though this network cannot meet all the needs alone:

> "In response to learning and innovation needs as a whole within an organization, they (the communities of practice) are making way for new collaboration models in the Internet age. This important technological development is linked to the knowledge economy, which requires increased collaboration in order to improve competitiveness on a larger scale." [26].

Due to the evolution of methods of communication, the healthcare system and its practitioners must adapt to this new reality in order to continue to provide quality care, and communities of practice such as the RIUPS are a promising way forward for patients with chronic diseases. Universities are falling into step [25], and since this is no longer a choice, in the short-term, healthcare organizations will have to adapt, spurred by the new generation raised with social media. These new technologies, which facilitate knowledge transfer, will allow healthcare professionals and patients who are part of the RIUPS to not only take advantage of a knowledge exchange bent on promoting best practices but also benefit from a network that fosters real collaboration. Overall, partners will save time while drawing participants' attention to their personal concerns, thus ensuring the results of these collaborations are better suited to the needs of each. The efforts and energy invested in developing the RIUPS will surely be outweighed by the positive effects, for both patients and professionals, which the network will have on healthcare practices.

Acknowledgments

We would like to thank the Canadian Institutes for Health Research (CIHR), the Health Research Foundation (HRF) and the Vice-President of Research at Université de Montréal for their financial support.

References

[1] Lefebvre, H., Brault, I., Levert, M.-J., Roy, O., Proulx, M., Alarie, S. and Larivière, M. (2015). Une innovation pour un meilleur continuum de soin en oncologie: développement et implantation d'un portail web d'échange des savoirs. *Revue francophone internationale de recherche infirmière*. Advanced online publication. Doi: 10.1016/j.refiri. 2015. 10. 003.

[2] Institut de la statistique du Québec. (2014). La lutte contre le cancer au Québec et en Ontario: *une approche centrée sur le patient*. Retrieved from http://www.santecom.qc.ca/Bibliothequevirtuelle/Isq/978255070 4454.pdf.

[3] Ministère de la Santé et des Soins de longue durée. (november 2011). Améliorer le continuum des soins: Rapport du Comité consultatif

concernant les hospitalisations évitables. Toronto : Gouvernement de l'Ontario. Imprimerie de la Reine. Retrieved from http:// www. health. gov.on.ca/fr/common/ministry/publications/reports/baker_2011/baker_2 011.pdf.

[4] Allaudeen, N., Vidyarthi, A., Maselli, J. and Auerbach, A. (2011). Redefining readmission risk factors for general medicine patients. *Journal of Hospital Medicine*, *6*(2), 54-60.

[5] Nekhlyudov, L. and Schnipper, J. L. (2012). Cancer survivorship care plans: What can be learned from hospital discharge summaries? *Journal of Oncology Practice*, *8*(1), 24-29.

[6] Naylor, M. D., Brooten, D., Campbell, R., Jacobsen, B. S., Mezey, M. D., Pauly, M. V. and Schwartz, J. S. (1999). Comprehensive discharge planning and home follow-up of hospitalized elders: a randomized clinical trial. *Jama*, *281*(7), 613-20.

[7] Pipe, T. B., Cisar, N. S., Caruso, E. and Wellik, K. E. (2008). Leadership strategies: Inspiring evidence-based practice at the individual, unit, and organizational levels. *Journal of Nursing Care Quality*, *23*(3), 265-271.

[8] Polit, D. F. and Beck, C. T. (2008). *Nursing research: Generating and assessing evidence for nursing practice* (8th ed.). Philadelphia: Williams and Wilkins and Wolters Kluwer Health.

[9] Wenger, E., McDermott, R. and Snyder, W. M. (2002). A guide to managing knowledge: Cultivating communities of practice. Boston: Harvard Business School Press.

[10] Prax, J. Y. (2015). Les communautés de pratiques. Retrieved from http://www.polia-consulting.com/Les-communautes-de-pratiques.html.

[11] Laferrière, T. (november 2015). La communauté de pratique : quoi et pourquoi ? Paper presented at la journée thématique : Les communautés de pratiques : partager efficacement pour mieux collaborer à la 19e Journée annuelle de santé publique, Montréal. Abstract retrieved from http://jasp.inspq.qc.ca/communautes-de-pratique.aspx.

[12] Davel, E. and Tremblay, D. G. (2011). Formation et apprentissage organisationnel: La vitalité de la pratique. Québec: Presses de l'Université du Québec.

[13] Gherardi, S., Nicolini, D. and Odella, F. (1998). Toward a social understanding of how people learn in organizations: a notion of situated curriculum. *Management Learning*, *29*(3), 273-297.

[14] Kietzmann, J., Plangger, K., Eaton, B., Heilgenberg, K., Pitt, L. and Berthon, P. (2013). Mobility at work: A typology of mobile communities of practice and contextual ambidexterity. Journal of

Strategic Information Systems, 3(4), 1-16. doi: 10.1016/j.jsis. 2013. 03. 003.

[15] Lave, J. and Wenger, E. (1991). *Situated learning: Legitimate peripheral participation.* New York: Cambridge University Press.

[16] Snyder, W. M. and Wenger, E. (2000). Cultiver vos réseaux invisibles. L' *Expansion Management Review*, march, 6-12.

[17] Wenger, E. (1998). *Communities of practice: Learning, meaning, and identity*. Cambridge: Cambridge University Press.

[18] Mitchell, J. (2002). The potential for communities of practice to underpin the national training framework. Melbourne: Australian National Training Authority.

[19] Hendry, C. (1996). Understanding and creating whole organizational change through learning theory. *Human Relations*, *49*(5), 621-641.

[20] Gherardi, S. and Nicolini, D. (2002). Learning the trade: a culture of safety in practice. *Organization*, *9*(2), 191-223.

[21] Meingan, D. (2002). Les communautés de pratique ou le partage de savoirs. *Expansion Management Review*. Septembre, 46-52.

[22] Bourhis, A. and Tremblay, D. G. (2004, juillet). Les facteurs organisationnels de succès des communautés de pratique virtuelles. Retrieved from http://www.cefrio.qc.ca/media/uploader/Facteurs_organisationnels.pdf.

[23] Chanal, V. (2000). Communautés de pratique et management par projet : en référence à Wenger (1998). Communities of practice. Learning, meaning, and identity. *Management*, *3*(1), 1-30.

[24] Commission européenne. (2001). Réaliser un espace européen de l'éducation tout au long de la vie. Retrieved from http://eur-lex. europa. eu/legal-content/FR/TXT/?uri=uriserv:c11054.

[25] Jalali, A. (november 2015). Pourquoi, quand et comment utiliser les médias sociaux à des fins pédagogiques? Communication présentée au colloque Enseigner à l'ère des médias sociaux: les avantages et les enjeux, Montréal.

[26] Laferrière, T., Milton, C., Benoit, J. (2004, june). *Apprendre au quotidien dans l'organisation québécoise : le cas des communautés de pratique virtuelles*. Montréal: Cefrio.

[27] Campos, M. N., Laferrière, T. and Harasim, L. (2001). The post-secondary networked classroom: renewal of teaching practices and social interaction. *Journal of Asynchronous Learning Networks*, *5*(2), 36-52. Retrieved from https://www.researchgate.net/publication/

228587797_The_post-secondary_networked_classroom_Renewal_of_teaching_practices_and_social_interaction.

[28] Campos, M. N. (2003). The progressive construction of communication: Towards a model of cognitive networked communication and knowledge communities. *Canadian Journal of Communication*, *28*(3), 291-322.

[29] Gordon, T. and Sterling, E. W. (1997). Making the patient your partner : communication skills for doctors and other caregivers (éditions logistiques). Wesport: Aubun House.

[30] Purtilo, R. B. (1978). Health professional/patient interaction: Allied health professional and the patient (2nd ed). Philadelphia: Saunders.

[31] Mercier, C. and Alarie, S. (2001). Le processus de rétablissement chez les personnes alcooliques et toxicomanes. In P. Brisson (Ed.), *L'usage de drogues et la toxicomanie*, (vol. *III*, pp. 335-365). Montréal: Gaëtan Morin Éditeur.

[32] Woods, S. S., Schwatz, E., Tuepker, A., Press, N. A., Nazi, K. M., Turvey, C. L. and Nichol, W. P. (2013). Patient experience with full electronic access to health records and clinical notes through the My Healthevet Personal Health Record Pilot: Qualitativ study. *Journal of Medical Internet Research*, *15*(3), e65. doi: 10.2196/jmir.2356.

[33] Larin, L. R. (2014). Inspired to change: improving patient care one story at a time. Chicago: ACHE management series.

[34] Lefebvre, H. (april 2015). Une recherche pour et avec les patients. Paper presented at la 2eme Journée Francophone de la recherche en soins. Recherche industrielle et académique : lien entre sciences et pratiques cliniques. Angers.

[35] Pomey, M. P. and al. (2015). Le «Montreal model»: enjeux du partenariat relationnel entre patients et professionnels de la santé. *Santé publique*, *27*(1), 41-50.

[36] Centre de pédagogie appliquée aux sciences de la santé (CPASS) de l'Université de Montréal. (2013, August 29). Guide d'implantation du partenariat de soins et de services, vers une collaboration optimale entre intervenants et avec le patient. Montréal: Université de Montréal, Direction collaboration et partenariat patient/CPASS, 2013a. Retrieved from http://ena.ruis.umontreal.ca/pluginfile.php/256/coursecat/description/Guide_implantation1.1.pdf.

[37] Richards, T., Montori, V. M., Godlee, F., Lapsley, P. and Paul, D. (2013). Let the patient revolution begin. *BM*, *346*, *f2614*, doi: http://dx.doi.org/ 10.1136/bmj.f2614.

[38] Karazivan, P., et al. (2015). The patient as partner in care: Conceptual framework for a necessary transition. *Academic Medicine*, *90*(4), 437-441. doi: 10.1097/ACM.0000000000000603.

[39] Coulter, A. (2002). The autonomous patient: Ending paternalism in medical care. *Journal of the Royal Society in Medicine*, *95*(12), 623-624. Retrieved from http:// www.ncbi.nlm.nih.gov/ pmc/ articles/ PMC1279 295/.

[40] Lorig, K. R. and al. (1999). Evidence suggesting that a chronic disease self-management program can improve health status while reducing hospitalization: a randomized trial. *Medical Care*, *37*(1), 5-14.

[41] The National health service (NHS). (2001). The expert patient: A new approach to chronic disease management for the 21st century. London: Department of health.

[42] Ministère de la Santé et des Services Sociaux. (2006). Direction de la lutte contre le cancer: Rapport d'activités 2005-2006. Québec: Gouvernement du Québec.

[43] Comité consultatif concernant les hospitalisations évitables. (2011). Améliorer le continuum de soins. Toronto: Ministère de la Santé et des Soins de longue durée.

[44] Wyer, P. C. (2007). Responsiveness to change: A quality indicator for assessment of knowledge translation systems. *Academic Emergency Medicine*, *14*(11), 928-931.

[45] Grol, R. and Buchan, H. (2006). Clinical guidelines: what can we do to increase their use? *Medical Journal of Australia*, *185*(6), 301-302.

[46] Parmelli, E., Flodgren, G., Beyer, F., Baillie, N., Schaafsma, M. E. and Eccles, M. P. (2011). The effectiveness of strategies to change organisational culture to improve healthcare performance: a systematic review. *Implementation Science*, *6* (33). doi:10.1186/1748-5908-6-33.

[47] Dopson, S., FitzGerald, L., Ferlie, E., Gabbay, J. and Locock, L. (2010). No magic targets! Changing clinical practice to become more evidence based. *Health Care Manage Review*, *35*(1), 2-12.

[48] Nelson, A. (september 2009). *Top ten reasons why programs fail… and what to do about it* in tranforming. Patient safety. Fall. Conference. University of South Florida. http://documents.mx/documents/patient-safety-2009-falls-conference.html.

[49] Estabrooks, C. A., Rutakumwa, W., O'Leary, K., Profetto-McGrath, J., Milner, M., Levers, M.-J. and Scott-Findlay, S. (2005). Sources of practice knowledge among nurses. *Qualitative Health Research*, *15*(4), 460-476.

[50] Graham, I. D., Logan, J., Harrison, M. B., Straus, S. E., Tetroe, J., Caswell, W. and Robinson, N. (2007). Lost in knowledge translation: Time for a map? *Journal of Continuing Education in the Health Professions*, *26*(1), 13-24.

Biographical Sketch

Hélène Lefebvre
Université de Montréal

Education: PhD, Université de Montréal

Address: Faculty of Nursing
Pavillon Marguerite-d'Youville
C.P. 6128 succ. Centre-ville
Montréal (Quebec) H3C 3J7CP

Research and Professional Experience:

Hélène Lefebvre is a full professor at Université de Montréal's Faculty of Nursing; a visiting professor at the School of Public Health of the Université libre de Bruxelles; a visiting professor at Université Aix-Marseille; and a visiting researcher at the research laboratory of the Faculty of Nursing at Paris University 13. She is co-director of the Interdisciplinary Research Group on Adaptation and Resilience (GIRR) at the Centre for Interdisciplinary Research in Rehabilitation of Greater Montreal (CRIR). She is also a member of the *Réseau de recherche en intervention en sciences infirmières du Québec* (RRISIQ) (Quebec Research Network on Nursing Interventions), as well as a researcher at the Quebec Rehabilitation Research Network (REPAR). She heads the *Réseau infirmier, un partenaire de soins* (RIUPS) (The nursing network: a partner in care), which promotes knowledge exchange to improve clinical practices and patient care. She is the principal investigator appointed by the CIHR for the project "Developing a clinical innovation in oncology: toward an improved continuum of care and services." She also conducts studies on the social participation and resilience of individuals with chronic diseases or disabilities, from an educational intervention perspective, through continuity of services and the community and social integration of these individuals. She is a leader in social participation, healthcare partnerships, patient partnerships, community integration intervention, and the exchange of knowledge developed through the relationship between nurses, interveners, managers, individuals and families who are partners in care in inter and cross-sectoral contexts.

Conferences

Levasseur, M., Lefebvre, H., Levert, M.-J., Lacasse-Bédard, J., Desrosiers, J., Therriault, P.-Y., Tourigny, A., Couturier, Y., Carbonneau, H. (2016). Personalized attendant for social participation (APIC) in older adults: a promising intervention to live longer and better? Congrès mondial sur la résilience. Université du Québec à Trois-Rivières (UQTR), Trois-Rivières, Quebec, Canada. August 22-24, 2016.

Swaine, B; Kehayia, E; Labbé, D; Poldma, T; Fung, J; Archambault, P; Kairy, D; Lamontagne, A; Le Dorze, G; Lefebvre, H; Ahmed, S. (2015). Partnering to support participation in accessible and inclusive environments using a Living Lab approach. Proceedings of the International Mobility Conference, July 6-10, Montreal, QC. p. 67.

Brault, I., Lefebvre, H., Levert, M.-J., Roy, O (2015). A Web Knowledge Exchange Portal for a Better Discharge Planning in Oncology. International Conference on Cancer Nursing (ICCN), July 8-12, 2015. Vancouver, British Columbia, Canada.

Lefebvre, H., Levert, M.-J. (May 2015) Les espaces publics promoteurs de l'autonomie et d'efficacité dans leur fréquentation? 1er congrès québécois de recherche en adaptation. *Rencontre des forces vives et de l'excellence en réadaptation physique*. May 21-22. Boucherville.

Lefebvre, H., Brault, I. Roy, O., Levert, M.-J. (May 2015). Réseau infirmier: un partenaire de soins. 6ème Congrès international du SIDIIEF. *Le défi des maladies chroniques : un appel à l'infirmière.* May 31. Montréal.

Gauvin-Lepage, J., Lefebvre, H. and Malo, D. (submitted September 2015). *Soutenir le processus de résilience familiale par l'utilisation de stratégies d'interventions interprofessionnelles : proposition d'un programme d'intervention novateur.* Congrès annuel de Lésion cérébrale Canada, Montréal, Quebec.

Gauvin-Lepage, J., Lefebvre, H. and Malo, D. (Guest speaker – September 2015). *Soutenir le processus de résilience familiale par l'utilisation de stratégies d'interventions interprofessionnelles : une étude issue du milieu de la réadaptation.* 33e journées d'études de l'Association des infirmières de rééducation et réadaptation, Poitiers, France, Europe.

Gauvin-Lepage, J., Lefebvre, H. and Malo, D. (accepted June 2015). *Soutenir le processus de résilience familiale par l'utilisation de stratégies d'interventions interprofessionnelles : une étude issue du milieu de la réadaptation.* 6e congrès mondial des infirmières et infirmiers francophones (SIDIIEF), Montréal, Quebec.

Gauvin-Lepage, J., Lefebvre, H. and Malo, D. (Guest speaker – February 2015). *Resilience in families of adolescents with traumatic brain injuries: development of a support intervention.* Interprovincial Student Journal Club. Canada.

Gauvin-Lepage, J., Lefebvre, H. and Malo, D. (Guest speaker – October 2014). *Soutenir la résilience de familles dont un adolescent est atteint d'un traumatisme craniocérébral : proposition d'un programme d'intervention novateur.* Colloque. Réadaptation et résilience : leçons de l'expérience, Trois-Rivières, Québec.

Gauvin-Lepage, J., Lefebvre, H. and Malo, D. (Guest speaker – October 2014). *La résilience familiale : considérer les forces inhérentes et le potentiel de grandir de la famille.* Rencontre clinique de la direction des services ambulatoires. Montreal Children's Hospital, Montréal, Quebec.

Gauvin-Lepage, J., Lefebvre, H. and Malo, D. (Guest speaker – May 2014) *Soutenir le processus de résilience familiale par l'utilisation de stratégies d'interventions interprofessionnelles : une étude issue du milieu de la réadaptation.* Nursing Grand Rounds of the Jewish Rehabilitation Hospital, Laval, Quebec.

Rothan-Tondeur, M., Lefebvre, H., Eymard, C. (2015) PIRI, Partenariat International pour la Recherche Infirmière. 6e Congrès mondial des infirmières et infirmiers francophones du Secrétariat International des Infirmières de l'Espace Francophone. May 30 to June 5, 2015. Montréal, Canada.

Gulpa N., Lefebvre, H., Roy, O. (2015). Intervention permettant une meilleure prise en charge des autosoins par la clientèle adulte de chirurgie intestinale. 6e congrès Mondial, *Défi des maladies chroniques : un appel à l'expertise infirmière. Secrétariat international des infirmières et infirmiers de l'espace francophone*. May 30 to June 5, 2015. Montréal.

Blain S, Lefebvre H, Levert M-J, Desrosiers J, Carbonneau H, Thérriault P-Y, Couturier Y, Larivière N and Levasseur M. (2014). Accompagnement Personnalisé d'intégration Communautaire (APIC) des personnes âgées en perte d'autonomie : résultats préliminaires. *10e Journée scientifique du 1er cycle de la FMSS de l'Université de Sherbrooke*. September 12.

Gauvin-Lepage, J., Lefebvre, H. and Malo, D. (2014). Soutenir la résilience de familles dont un adolescent est atteint d'un traumatisme crâniocérébral en soutien à la résilience. REPAR. May.

Gauvin, Lepage, J., Lefebvre, H. and Malo, D. (2014) Soutenir la résilience de familles dont un adolescent est atteint d'un traumatisme crâniocérébral : proposition d'un programme d'intervention novateur 2e Congrès mondial sur la résilience. *De la personne à la société.* Timisoara, Romania. May.

Brault, I., Lefebvre, H. and Roy, O. (2013). *La planification du congé en oncologie : le rôle du Portail d'échanges des savoirs.* Congrès ADELF-SFSP Santé Publique et Prévention, Bordeaux, France. October.

Gauvin-Lepage, J., Lefebvre, H. and Malo, D. (2013). *Resilience in families of adolescents with traumatic brain injuries: development of a support intervention.* 90th Annual American Congress of Rehabilitation Medicine Conference, Progress in Rehabilitation Research, Orlando, Florida. November.

Levasseur, M., Lefebvre H. (2013) Un accompagnement personnalisé en soutien à la participation sociale d'aînés en perte d'autonomie : validation et développement d'un partenariat autoportant. *81e Congrès de l'Association francophone pour le savoir (ACFAS) – Colloque "La participation sociale des aînés : diversités, égalités et solidarité".* May 6-10. Quebec.

Lefebvre, H. and Levert, M.-J. (2013). Un accompagnement personnalisé dans la communauté peut-il soutenir la résilience des personnes ayant des incapacités? Colloque International sur la résilience, ACFAS, May 6-10. Quebec.

Levert, M.-J., Lefebvre, H., Gélinas, I., Croteau, C., LeDorze, G., Samuelson, J., Perroux, M., Vincent, P. and Mitchell, P. (2013). *Accompagnement-citoyen personnalisé d'intégration communautaire (APIC) – Tuteur de résilience.* Brain Injury Association of Canada, Kingston. September 25-27.

Michallet, B. and Lefebvre, H. (November 2013). *Les défis de la participation sociale des aînés ayant des incapacités : Interagir pour des politiques, des recherches et des pratiques porteuses de changement.* Colloque "Les défis de la participation sociale des aînés ayant des incapacités: Interagir pour des politiques, des recherches et des pratiques porteuses de changement", RIPPH. Quebec.

Lefebvre, H., Levert, M.-J., Perroux, M., Le Dorze, G., Croteau, C., Therriault, P.-Y. and Michallet, B. (2013, mai). *Que disent les personnes ayant un traumatisme crânien de leur fréquentation des lieux publics?* Colloque Innovation et recherche en santé : l'approche Living Lab, ACFAS, Quebec.

Michallet, B., Lefebvre, H. and Chouinard, J. (May 2013). *Résilience et réadaptation en déficience physique; pour un renouvellement des pratiques cliniques et organisationnelles.* Colloque International sur la résilience, ACFAS, Quebec.

Kehayia, E., Swaine, B, Ahmed, S., Archambault, P., Fung, J., Lamontagne, A., Le Dorze, G., Lefebvre, H., Longo, C., Overbury, O. and Poldma, T. (April 2013). *From rehabilitation to participation and inclusion of people with disabilities: creating and transforming environments for all.* The Pacific Rim International Conference on Disability and Diversity, Hawaii.

Berger, V., Eymard, C., Lefebvre, H. (2013). *Stratégie d'implantation d'un outil pédagogique et didactique d'évaluation du risque de constipation du patient hospitalisé : Exemple d'une recherche collaborative.* Apprentissage, Didactique, Evaluation, Formation, Université de Provence, en association avec l'Institut National de Recherche Pédagogique, l'Unité Mixte de Recherche P3.

Levert, M.-J., Perroux, M., Le Dorze, G., Croteau, C., Therriault, P.-Y. and Michallet, B. (April 2013). *Que disent les personnes ayant un traumatisme crânien de leur fréquentation des lieux publics?* Poster. Carrefour des connaissances, Centre de réadaptation Lucie-Bruneau, Montréal. April 9.

Levert, M.-J., Lefebvre, H., Perroux, M., Le Dorze, G., Croteau, C., Therriault, P.-Y. and Michallet, B. (April 2013). *Accompagnement personnalisé en intégration communautaire en soutien à la résilience.* Carrefour des connaissances, Centre de réadaptation Lucie-Bruneau, Montréal. April 9.

Michallet, B., Lefebvre, H. and Chouinard, J. (October 2012). *Résilience: pour un renouvellement des pratiques cliniques et organisationnelles en réadaptation.* 11th Congress of World Association for Psychosocial Rehabilitation, Milan, November 10-13, 2012.

Malo, D., Levert, M.-J., Levasseur, I., Lefebvre, H., McKerral, M., Roy, O., Thériault, P.-Y., Guindon, A. and Dostie, M. (October 2012). *Préserver l'autonomie fonctionnelle des personnes âgées ayant subi un traumatisme cranio-cérébral léger (TCCL) : Adaptation, implantation et évaluation de l'intervention d'Accompagnement personnalisé d'intégration communautaire (APIC).* Présentation par affiche. Colloque «Vieillir au XXIe siècle : approches préventives, diagnostiques et restauratives», Montréal, Canada.

Malo, D., Levert, M.-J., Levasseur, I., Lefebvre, H., McKerral, M., Roy, O., Thériault, P.-Y., Guindon, A. and Dostie, M. (November 2012). *Preventing functional decline of frail elders following mild traumatic brain injury*. Poster presentation. 11th World Congress of the World Association of Psychosocial rehabilitation, Milan, Italy.

Lefebvre, H., Levert, M.-J., Bottari, C., Gélinas, I., Croteau, C., Le Dorze, G., Samuelson, J., Perroux, M. Vincent, P. and Mitche, P. (October 2012). *Accompagnement personnalisé d'intégration communautaire.* Poster presentation. Journée Annuelle de Santé Publique, Montréal.

Rochette, A., Racine, E., Lefebvre, H., Lacombe, J., Bastien, J. and Tellier, M. (October 2012). *Systematically including relatives as clients into the rehabilitation process after stroke: related ethical issues perceived by relatives, patients and health professionals.* Canadian Stroke Congress, Calgary.

Gauvin-Lepage, J. and Lefebvre, H. (October 2012). *Résilience de familles dont un adolescent est atteint d'un traumatisme craniocérébral : développement d'une intervention de soutien.* Brain Injury Association of Canada.

Boyer, L., Tardif, J. and Lefebvre, H. (June 2012). A Collaborative Research on the Evaluation of Nursing Competencies. *The 4th International Nurse Education Conference Changing the landscape for nursing and healthcare education: evidence-based innovation, policy and practice.* Baltimore. June 19.

Boyer, L., Tardif, J., Lefebvre, H., (May 2012). *Des critères congruents avec l'évaluation du développement des compétences infirmières*, Congrès international du SIDIIEF 2012, Geneva, Switzerland, May 22.

Boyer, L., Tardif, J., Lefebvre, H (May 2012). *Des cibles de l'apprentissage cohérentes avec le concept de compétence : Une prémisse pour la validité de l'évaluation des compétences*, Congrès de l'Association internationale de pédagogie universitaire (AIPU), Trois-Rivières, Quebec, May14.

Lefebvre, H. (June 2012). *Résilience et Accompagnement citoyen Personnalisé en Intégration dans la communauté : résultats d'une étude longitudinale.* 1er Congrès Mondial sur la Résilience De la recherche à la pratique, Paris.

Thériault, P.-Y., Lefebvre, H., Levert, M.-J. and Briand, C. (June 2012). *Mieux accompagner des accompagnateurs-citoyens par l'approche de la psychodynamique du travail : un pas de plus pour supporter la résilience.* 1er Congrès Mondial sur la Résilience De la recherche à la pratique, Paris.

Gauvin-Lepage, J. and Lefebvre, H. (May 2012). *Resilience in families of adolescents with severe traumatic brain injuries: development of a support intervention.* American Family Therapy Academy Annual Conference, Theme: Family Resilience, San Francisco, US.

Soulard, N. and Lefebvre, H. (May 2012) *Le développement et l'implantation d'un outil d'évaluation de la pratique professionnelle des infirmières débutantes en milieu clinique. V^e^ congrès mondial des infirmières et infirmiers francophones (SIDIIEF)*, Geneva, Switzerland.

Lefebvre, H. (May 2012). *Accompagnement citoyen d'intégration dans la communauté en soutien à la résilience des personnes ayant des incapacités.* V^e^ congrès mondial des infirmières et infirmiers francophones (SIDIIEF), Geneva, Switzerland. Retrieved: http:// www.sidiief.org/~/ media/Files/ 7_0_Publications/7_1_ PublicationsSIDIIEF/7_1_9_PresentationsCongresMondiaux/7_1_9_3_PresentationsCongres2012/SP117.4-Lefebvre-Helene.ashx

Lefebvre, H., Roy, O., Dumais, R. and Lanctôt, L. (May 2012). *Réseau infirmier du RUIS de l'Université de Montréal: l'échange des savoirs dans une approche collaborative.* V^e^ congrès mondial des infirmières et infirmiers francophones (SIDIIEF), Geneva, Switzerland.

Gauvin-Lepage, J. and Lefebvre, H. (March 2012). *Resilience of families who have a teenager with severe head injuries: Development of an intervention of support.* 9th World Congress on Brain Injury. International Brain Injury Association, Edinburgh, Scotland.

Rochette, A., Racine, E., Lefebvre, H., Lacombe, J., Bastien, J. and Tellier, M. (March 2012). *Systematically including relatives as clients into the rehabilitation process after stroke: related ethical issues perceived by relatives, patients and health professionals*. Canadian Stroke Congress, Calgary.

Lefebvre, H., Levert, M.-J., Botari, C., Gélinas, I., Croteau, C., Le Dorze, G., Therriault, P.-Y., Samuelson, J. and Perroux, M. (March 2012). *Accompagnement citoyen personnalisé d'intégration communautaire : Tuteur de résilience?* Poster presentation. Carrefour des connaissances, Centre de réadaptation Lucie-Bruneau, Montréal.

Therriault, P.-Y., Lefebvre, H., Levert, M.-J., Béland, C. and Bélec-Dupuis, M. (March 2012). *S'intéresser aux accompagnateurs pour mieux soutenir le processus d'accompagnement des personnes ayant un TCC sévère.* Poster presentation. Carrefour des connaissances. Centre de réadaptation Lucie-Bruneau, Montréal.

Invited Conferences

Lefebvre, H. (2015). Enjeux liés au transfert des connaissances. *Groupe de contact en sciences infirmières.* November 30. Brussels.

Lefebvre, H. (2015). Modèle humaniste des soins infirmiers – UdeM-Inspiration pour la pratique. CHU Charleroi. December 1st.

Lefebvre, H. (2015). Le Réseau infirmier un partenaire de soins. Université Libre de Bruxelles, Erasme. December 2.

Lefebvre, H. (2015). Modèle humaniste des soins infirmiers – UdeM-Inspiration pour la pratique. Université Libre de Bruxelles, Erasme. December 3.

Lefebvre, H. (2015). Pourquoi est-ce important pour les patients de participer en tant que partenaire de la formation, de la clinique et de la recherche. *5e Colloque inter-régional Recherche paramédical grand Sud-Ouest. Bordeaux*. September 24-25.

Lefebvre, H. (2015). L'Evidence based practice : un lien entre la pratique, la formation et la recherche. 2eme Journées Francophone de la recherche en soins. *Recherche industrielle et académique : lien entre sciences et pratiques cliniques.* Angers. France, April 9-10.

Lefebvre, H. (2015). Une recherche pour et avec les patients. 2eme Journées Francophone de la recherche en soins. *Recherche industrielle et académique : lien entre sciences et pratiques cliniques.* Angers. France, April 9-10.

Lefebvre, H. (2015). Transfert de connaissances en co-construction avec les milieux cliniques. *1ere Journée scientifique sur la recherche en science infirmière et paramédicale : Enjeux et perspectives.* Marseille, France. April 21.

Lefebvre, H. (2015). Enjeux et défis du partenariat de soins. Atelier Partenariat de soins. Les patients partenaires : partage d'expériences et mise en réseau des pratiques. *1ere Journée scientifique sur la recherche en science infirmière et paramédicale : Enjeux et perspectives.* Marseille, France. April 21.

Lefebvre, H. (2015). Le patient partenaire dans la recherche, la formation et le soin. *1ere Journée scientifique sur la recherche en science infirmière et paramédicale : Enjeux et perspectives.* Marseille, France. April 21.

Lefebvre, H., Berger, V., Eymard, C. (2015). Un Portail d'Échanges et de Savoirs (PES) en soutien aux équipes cliniques en oncologie : Une innovation clinique pour le transfert des connaissances entre le Québec et la France. Bordeaux. January 27.

Cara, C. Lefebvre H. (2015). Le Modèle humaniste UdeM, une inspiration pour la clinique. Brussels. February 3. Belgium.

Cara, C. Lefebvre H. (2015). Le Modèle humaniste UdeM, une inspiration pour la gestion. Brussels. February 4. Belgium.

Cara, C. Lefebvre H. (2015). Le Modèle humaniste UdeM, une inspiration pour la formation. Brussels. February 5-6. Belgium.

Lefebvre, H. (2014). Modèle humaniste de soins-UdeM. Préparé en collaboration avec Chantal Cara. École d'été, Institut de santé publique, épidémiologie et développement, Université de Bordeaux. June 5.

Lefebvre, H. (2014). Épistémologie de sciences infirmières et modèle humaniste. Conférence publique. DIU Faculté de médecine, Université Aix-Marseille. December 10. Marseille.

Lefebvre, H. (2014). Pratique infirmière avancée. Conférence publique. DIU Faculté de médecine, Université Aix-Marseille. December 10. Marseille.

Lefebvre, H. (2014). Apport de la recherche infirmière en cancérologie *Rencontres de la cancérologie française 2014*. December 3. Paris.

Lefebvre, H. (2014). Le transfert de connaissances en recherche infirmière. *Troisième journée APHP sur la recherche infirmière et paramédicale"* Hôpital Pitié Salpêtrière. Assistance Publique Hôpitaux de Paris. November 28. Paris.

Lefebvre, H. (2014). La place de l'infirmière clinicienne au sein du système de santé et ses interactions: Point de vue Canadien. *Fédération nationale des étudiants en soins Infirmiers (FNESI).* November 22. Nantes.

Lefebvre, H. (2014). Partenariat de soins. Conférence publique. DIU Faculté de médecine, Université Aix-Marseille. November 17. Marseille.

Lefebvre, H. (2014). Épistémologie de sciences infirmières et modèle humaniste. Conférence publique. DIU Faculté de médecine, Université Aix-Marseille. November 24. Marseille.

Lefebvre, H. (2014). *Planification de congé et transitions des soins en oncologie*. 1ère journée scientifiques *Réseau infirmier un partenaire de soins* Faculté des sciences infirmières Université de Montréal. October 15. Montréal.

Lefebvre, H. (2014). La pratique infirmière avancée. Conférence publique. DIU Faculté de médecine, Université Aix-Marseille. April. Marseille.

Lefebvre, H. (2014) Un accompagnement citoyen personnalisé en intégration communautaire en soutien à la résilience. Conférence publique. École doctorale École des hautes Études en Santé Publique. March. Rennes.

Lefebvre, H. (2014) Quand la recherche et la clinique vont de pair: Illustration d'un partenariat entre la recherche, la clinique et la communauté. Conférence publique. École de Santé Publique, Université Libre de Bruxelles. March. Brussels.

Lefebvre, H. (2014) Proposition d'un cadre de référence : Le modèle humaniste des soins UdeM. Association internationale de recherche en sciences infirmières. April. Paris.

Lefebvre, H. (2013). Vers un partenariat de formation et de recherche en sciences infirmières : Pour le développement continu des compétences. Conférence publique. Université Libre de Bruxelles. November. Brussels.

Lefebvre, H. (2013) Le transfert des connaissances infirmières dans le milieu clinique. 1res Journées Francophones de la Recherche en Soins. April 11-12. Angers, France.

Lefebvre, H. (2013) Lien entre recherche et pratique. Journée d'études de l'Association de recherche en soins infirmiers. Paris, January 25.

Lefebvre, H. (2013). Quand la recherche rime avec la clinique : Quinze ans de partenariat entre la recherche, la clinique et la communauté. Conférence inaugural. EHESP Paris/Rennes. January 22.

Lefebvre, H., Levert, M.-J. (2012) Un Accompagnement citoyen personnalisé peut-il soutenir la résilience des personnes ayant un traumatisme crânien? Colloque de l'Association québécoise des traumatisés crâniens. Montréal, November 22-23.

Lefebvre, H. (2012). Comment développer un partenariat entre équipes de soins, patients et familles? *Les rencontres interdisciplinaires du comité d'éthique : Mieux communiquer pour prévenir les conflits.* Hôpital du Sacré-Cœur de Montréal. Montréal, November 15.

Lefebvre, H. (2012) La recherche en sciences infirmières vers l'amélioration des pratiques cliniques et un soin aux patients qui fait la différence. (2012) 2 ième Colloque inter-régional. Recherche paramédicale. Évolution de la recherche en soins. Bordeaux. September 28.

Lefebvre, H. (2012) Rôle et importance de la recherche pour les pratiques en soins infirmiers. XXXIX Congrès Sniil. October 2-4, 2012. Montréal.

Lefebvre, H. (2012). Modèle humaniste de soins-UdeM. In collaboration with Chantal Cara. École d'été, Institut de santé publique, épidémiologie et développement, Université de Bordeaux. June 5.

Lefebvre, H., (2012). Quand pratique clinique et recherche se marient par la voie d'un partenariat personnes, familles, cliniciens et chercheurs Réaliste ou utopique? École d'été, Institut de santé publique, épidémiologie et développement, Université de Bordeaux. June 5.

Lefebvre, H. (2012) Réseau infirmier-Partenaires de soins. École d'été, Institut de santé publique, épidémiologie et développement, Université de Bordeaux. June 5.

Research projects

2013-2014 Networking grant

Project title: *Internationalisation des programmes et de la recherche à la Faculté des sciences infirmières*

Applicant: Hélène Lefebvre

Co-applicant: Francine Girard

Organization: DRI ($15,000)

2013-2016 CIHR PHSI Operating grant

Project title: *Developing a clinical innovation in oncology: for a better continuum of care and services for cancer patients*

Principal investigator: Hélène Lefebvre

Co-investigators: Isabelle Brault, Damien Contandriopoulos, Sylvie Lemay, Marie-Josée Levert, Francine Girard, Caroline Larue and Januz Kaczorowski

Decision-makers: Odette Roy and Sylvie Dubois

Organization: CIHR-PHSI ($291,916); FRS ($291,916); FRQS ($80,000)

2013-2016 Grant program: *Partnership for social development*

Project title: *Pour une offre régionale en accompagnement de PIMO*

Coordinators: Marie-Josée Deit and Marie-Josée Vaillancourt (dir. PIMO)

Co-investigators: Hélène Lefebvre and Marie-Josée Levert

Organizations: Human Resources and Skills Development Canada – HRSDC ($617,000); Centraide, for one year ($25,000); *Québec Ami des Aînés: Soutien aux actions communautaires*, for one year ($100,000)

2013-2015 QNRA grant

Project title: *Soutien à la participation sociale d'aînés en perte d'autonomie: validation auprès d'organismes communautaires et développement d'un partenariat autoportant.*

Coordinators: Mélanie Levasseur and Hélène Lefebvre

Co-investigators: Denise Malo, Andrée Sévigny, Hélène Carbonneau and Marie-Josée Levert

Potential collaborators: Ginette Aubin, Isabelle Wallach, Noémie Royer, Julie Beauchamp, Marie-Eve Bédard and Julie Castonguay

Organization: QNRA ($10,000)

2013-2014 CIHR Catalyst Grant

Project title: *Adaptation et validation d'une intervention d'Accompagnement Personnalisé d'Intégration Communautaire (APIC): vers une optimisation des services offerts pour répondre aux besoins de participation sociale des personnes âgées en perte d'autonomie*

Principal investigator: Mélanie Levasseur

Co-researcher: Hélène Carbonneau, Yves Couturier, Johanne Desrosiers, Nadine Larivière, Hélène Lefebvre, Denise Malo, Pierre-Yves Therriault, Émilie Raymond, Andrée Sévigny, André Tourigny, Serge Émond and Marie-Josée Levert

Organization: CIHR ($100,000)

2012-2016 CRIR research centre grant

Coordinators: Eva Kehayia and Bonnie Swaine

Program title: A Rehabilitation Living Lab: Creating Enabling Physical and Social Environments to Optimize Social Inclusion and Social Participation of Persons with Physical Disabilities

Co-investigators: Members of the CRIR including Hélène Lefebvre

Organization: FRSQ ($568,857 per year for four years)

2012-2014 Quebec Nursing Intervention Research Network grant

Principal investigators: Sylvie Cossette and Nancy Feely.

Organization: FRSQ ($515,000)

2011-2013 Operating grant

Project title: *Préserver l'autonomie fonctionnelle des personnes âgées ayant subi un traumatisme craniocérébral léger (TCCL): Adaptation, implantation et évaluation de l'intervention d'Accompagnement personnalisé d'intégration communautaire (APIC).*

Coordinator: Denise Malo

Co-investigators: Hélène Lefebvre, Isabelle Gélinas, Marie-Josée Levert and Michelle McKerral

Organization: FRSQ ($150,000)

2012-2013 Project grant

Project title: *Communautés de pratique: transfert de connaissance et innovation*

Coordinator: Dahlia Kayri

Co-investigators: Hélène Lefebvre, Guylaine LeDorze and Lise Poissant

Organization: MSSS ($80,000)

2011-2013 Operating grant

Project title: *Évaluation d'une intervention d'accompagnement personnalisé en intégration dans la communauté pour les personnes ayant un traumatisme craniocérébral*

Coordinator: Hélène Lefebvre

Co-investigators: Carolina Bottari, Claire Croteau, Isabelle Gélinas, Guylaine Le Dorze, McKerral Michelle, Luc Noreau and Pierre-Yves Therriault

Collaborators: Jérôme Gauvin-Lepage, Jacques Drolet and Nathalie Boucher

Organization: CIHR ($120,783) MOP 114933

2011-2013 Operating grant
Project title: *Évaluation d'une intervention d'accompagnement personnalisé en intégration dans la communauté pour les personnes ayant un traumatisme cranio-cérébral*
Coordinator: Hélène Lefebvre
Co-investigators: Carolina Bottari, Claire Croteau, Isabelle Gélinas, Guylaine Le Dorze, McKerral Michelle, Sylvie Nouseux, Luc Noreau and Pierre-Yves Therriault
Collaborators: Jérôme Gauvin-Lepage, Jacques Drolet and Nathalie Boucher
Organization: CIHR ($71,912) MOP110933

Professional Appointments:
Tenured professor, Faculty of Nursing, Université de Montréal

Honors:

People's Choice Poster Award: Brault, I., Lefebvre, H., Levert, M-J, Roy, O (2015). A Web Knowledge Exchange Portal for Better Discharge Planning in Oncology. International Conference on Cancer Nursing (ICCN), July 8-12, 2015. Vancouver, British Columbia, Canada.

Florence Prize - Rayonnement International, *Ordre des Infirmières et infirmiers du Québec*

Finalist - Women of Distinction Award YWCA.

David Strauss Award ACRM-BI-ISIG, *projet accompagnement personnalisé d'intégration dans la communauté pour les personnes ayant un traumatisme crânien.*

Provost Award for Excellence, *Innovation pédagogique en collaboration*, with Paul Lebel, Marie-Claude Vanier, Pierre-Yves Thériault and Richard Ratelle ($3,000)

Contest award for clinicians with the best poster presentation. Grisé, M.-C., Lefebvre, H. and Trempe, C. (April 12, 2007). *Intégration de la visioconférence dans le processus de transfert inter établissements – Perception des personnes ayant une lésion médullaire, des proches et des intervenants*. Poster presentation. Journée scientifique du REPAR, Montréal, Canada.

Excellence in Research Award - Canadian Association of Schools of Nursing (CASN)

Contribution Award for Excellence – Quebec Public Service for the continuum of care in traumatology of the *Institut d'administration publique de Québec*.

Contribution Award - Quebec Minister of Education for the book: *L'examen clinique dans la pratique infirmière* (Brûlé, Cloutier in collaboration Doyon)

Contribution Award for Excellence - *Canadian Journal of Education, 16*(3).

Publications Last 3 Years:

Books and research studies

Lefebvre, H. and Michallet, B. (dir.) (2011). Résilience: pour voir autrement l'intervention de réadaptation. *Revue développement humain, handicap et changement social, 19*(1). 214 pages.

Printed manuals and research reports

Lefebvre, H, Levert, M.-J., Gélinas, I., Croteau, C, Le Dorze, G. and Therriault, Y. (2015). Évaluation d'une intervention d'accompagnement personnalisé en intégration dans la communauté pour les personnes ayant un traumatisme craniocérébral. CIHR Report. 40 pages.

Lefebvre, H, Levert, M.-J., Gélinas, I., Croteau, C, Le Dorze, G. and Therriault, Y. (2014) Effets d'un d'accompagnement à l'intégration communautaire sur le devenir de personnes ayant un traumatisme crânien modéré ou sévère en post réadaptation. CIHR Report. 35 pages.

Lefebvre, H., Levert, M.-J., LeDorze, G., Croteau, C. and Therriault, P.-Y. (2013). *Un accompagnement citoyen personnalisé auprès de personnes ayant subi un traumatisme craniocérébral.* 45 pages. CIHR 2011-2013.

Beauregard, L., Guindon, A., Noreau, L., Boucher, N. and Lefebvre, H. (2013) *Le retour à la communauté après une lésion médullaire : l'expérience des personnes et leurs proches.* 77 pages. FRSQ 2010-2012

Book chapters

Gauvin-Lepage, G. and Lefebvre, H. (2015). Le contexte. In Rothan-Tondeur M. *Recherche infirmière : guide pour la rédaction d'un protocole de recherche*. Maloine éd.

Gauvin-Lepage, G. and Lefebvre, H. (2015). La problématique. In Rothan-Tondeur M. *Recherche infirmière : guide pour la rédaction d'un protocole de recherche*. Maloine éd.

Gauvin-Lepage, G. and Lefebvre, H. (2015). Les questions et les hypothèses de recherche. In Rothan-Tondeur M. *Recherche infirmière : guide pour la rédaction d'un protocole de recherche*. Maloine éd.

Gauvin-Lepage, G. and Lefebvre, H. (2015). Le but et les objectifs de recherche. In Rothan-Tondeur M. *Recherche infirmière : guide pour la rédaction d'un protocole de recherche*. Maloine éd.

Gauvin-Lepage, G. and Lefebvre, H. (2015). La justification. In Rothan-Tondeur M. *Recherche infirmière : guide pour la rédaction d'un protocole de recherche*. Maloine éd.

Gauvin-Lepage, J., Lefebvre, H. and Malo, D. (2015). Résilience de familles dont un adolescent est atteint d'un traumatisme craniocrébral : développement d'une intervention de soutien. In F. Julien-Gauthier et C. Jourdan-Ionescu (sous la dir.), *Résilience assistée, réussite éducative et réadaptation* (pp. 93-97). Québec, Quebec: CRIRES books online. Available at URL: http://lel.crires.ulaval.ca/public/resilience.pdf

Bertin, S., Longo, C., Poldma, T., Kehayia, E., Swaine, B., Ahmed, S., Archambault, P., Fung, J., Lamontagne, A., Lefebvre, H., Le Dorze, G. and Overbury, O. (2013). *Diversity as a catalyst for co-constructing knowledge holistically: Transforming experiences within an innovative research Living Lab known as MALL* (Mall As Living Lab). 17 pages.

Research publications or scholarly works in peer reviewed journals

Boyer, L., Tardif, J, Lefebvre, H. (2015). From a medical problem to a health experience: How nursing students think in clinical situations. *Journal of Nursing Education*, *54*(11). 625-632.

Lefebvre, H., Brault, I., Levert, M.-J., Roy, O., Proulx, M., Alarie, S., Larivière, S. in collaboration. Beauchamps, J., Carignan, M., Compagna, L., Déry, J., Leboeur, I., Tremblay, C. (2015). Le développement d'une innovation clinique en oncologie: pour un meilleur continuum de soins et de services pour les patients atteints de cancer. *Revue francophone internationale de recherche infirmière*. http://www.em-premium.com/article/1017666

Gauvin-Lepage, J., Lefebvre, H. and Malo, D. (2015). Family resilience following a physical trauma and efficient support interventions: A critical literature review. *Journal of Rehabilitation*, *81*(3), 34-42.

Van Schingen, É., Ladegaillerie, G., Lefebvre, H., Challier, M. P. and Rothan-Tondeur, M. (2015). La mobilité professionnelle des infirmières hospitalières. *Soins*, *60*(798), 25-30.

Gauvin-Lepage, J., Lefebvre, H. and Malo, D. (2015). Resilience in families with adolescents suffering from traumatic brain injuries. *Rehabilitation Nursing.0.* 1-9.

Berger, V., Eymard, C., Lefebvre, H. (2015). Stratégie d'implantation d'une échelle d'évaluation du risque de constipation : Un modèle de transfert des connaissances. *Revue francophone International de recherche infirmière. 1(01). 11-22.*

Gauvin-Lepage, J., Lefebvre, H. and Malo, D. (2014). Family resilience: Defining the concept from a humanist perspective. *Interdisciplinary Journal of Family Studies. XX*(2).22-36

Bérubé, M., Gélinas, C., Bernard, F., Gagné, A., Laizner, A. and Lefebvre, H. (2014). Evaluation of the feasibility and acceptability of a nursing intervention program aiming to facilitate the transition of SCI patients and their family from ICU to a trauma unit. *International Journal of Orthopaedic and Trauma Nursing 18*(4). 214-226.

Lefebvre, H. (2014). L'infirmière cadre intermédiaire : rôle stratégique dans l'évolution des pratiques cliniques. *Revue Soins Cadres*. No. 785. 20-22.

Lefebvre, H., (2014). Struggles of people with traumatic brain injury in public places. *ALTER, European Journal of Disability Research 8* (2014) 183–193. www.sciencedirect.com/science/article/pii/S187506721400039X

Rochette A., Racine E., Lefebvre, H., Lacombe J., Bastien J., Tellier M. (2014) Actual and ideal services in acute care and rehabilitation for relatives post-stroke from three perspectives: relatives, stroke-client and health professionals. Patient Educ Couns.___94(3):384-9. doi: 10.1016/j.pec.2013.10.028. Epub Nov. 5, 2013.

Vanier, M. C., Therriault, P.-Y., Lebel, P., Nolin, F., Lefebvre, H., Brault, I., Drouin, E. and Fernandez, N. (2014). Innovating in teaching of collaborative practice with a large Student cohort at Université de Montréal. *J Allied Health.* 42(4), e97-e106.

Michallet, B., Lefebvre, H., Hamelin, A. and Chouinard, J. (2014). Résilience et réadaptation en déficience physique : proposition d'une définition et éléments de réflexion. Resilience and rehabilitation in physical disability: Proposal for a definition and elements of reflection. *Revue Québécoise de Psychologie.* 35 : 1, 163-182.

Perroux, M., Lefebvre, H., Malo, D. and Levert, M.-J. (2013). Qu'en est-il des besoins des personnes ayant un traumatisme cranio-cérébral léger avec des séquelles persistantes? *Revue de santé publique. 25*(6). 719-728. http://www.cairn.info/revue-sante-publique-2013-6-page-719.htm

Berger, V., Eymard, C. and Lefebvre, H. (2013). Enjeux éthiques de l'infirmière dans la prévention de la constipation des patients hospitalisés : honte, peur et d'intimité. *Éthique et santé, 10*, 216-220.

Lefebvre, H., Levert, M.-J., Le Dorze. G., Croteau, C., Gélinas, I., Therriault, P.-Y., Michallet, B. and Samuelson, J. (2013). Un accompagnement citoyen personnalisé en soutien à l'intégration communautaire des personnes ayant subi un traumatisme craniocérébral : vers la résilience ? *Recherche en soins infirmiers*, *115*, 105-121.

Lefebvre, H., Roy, O., Satahli, S. and Rothan-Tondeur, M. (2013). Liens entre recherche et pratique : Échange des savoirs, utilisation des données probantes et renouvellement des pratiques. *Revue recherche en soins infirmiers*, 114, 31-45.

Levert, M.-J. and Lefebvre, H. (2013). Les technologies de la communication au service des personnes vivant avec une maladie chronique. *Point*, 20-23.

Rochette, A., Racine, E., Lefebvre, H., Lacombe, J., Bastien, J. and Tellier, M. (2013). Actual and ideal services in acute care and rehabilitation for relatives post-stroke from three perspectives: relatives, stroke clients and health professionals. *Journal of Rehabilitation Medicine*, *46*(1), 16-22.

Therriault, P.-Y., Lefebvre, H., Guindon, A., Levert, M.-J, and Briand, C. (2013). Accompagnement citoyen personnalisé en intégration communautaire : un défi pour la santé mentale? *Revue santé mentale au Québec*, *38*(1), 165-188.

Beauregard, L., Guindon, A., Noreau, L, Lefebvre, H. and Boucher, N. (2012). Community Needs of People Living With Spinal Cord Injury and Their Family. *Spinal Cord Injury Rehabilitation. 18*(2), 112-125.

Lefebvre, H. and Levert, M.-J. (2012). Services vs. perceived needs: A status report in Québec and France on persons with moderate or severe traumatic brain injuries. *Journal of Trauma Nursing*, *19*(4), 197-207.

Lefebvre, H. and Levert, M.-J. (2012). Les proches de personnes ayant un traumatisme crânien ont-ils des besoins particuliers? *Développement humain, handicap et changement social.* (Journal of Human Development, Disability and Social Change). *«Milieux urbains, politiques municipales et personnes ayant des incapacités»*, *19*(3), 87-100.

Lefebvre, H. and Levert, M.-J. (2012). Needs experienced and services according to mild brain injury patients and families. *Revista de Enferm Universide Federal do Piaui I. Brasila*, *1*(2), 128-134.

Lefebvre, H. and Levert, M.-J. (2012). The close relatives of people who have had a traumatic brain injury and their special needs. *Brain Injury*, *26*(9), 1084-1097.

Gauvin-Lepage, J., Lefebvre, H. and Noiseux, S. (2011) Intervention en soutien à la résilience familiale : une recension des écrits. Résilience :

pour voir autrement l'intervention en réadaptation. *Développement humain Handicap et changement social, 19*(1), 153-160.

Maryse Larivière
Université de Montréal

Education: M. A. Sociology, B. Sociology, C. Social Sciences

Address: Faculty of Nursing
Université de Montréal
CP. 6128 Succ. Centre-Ville
Montréal, Quebec Canada H3T 3J7

Research and Professional Experience:

Economy, sociology of work and organizations, education, personnel management and HR challenges, company training and assessment in various industries, communities of practice in knowledge-intensive sectors, industrial clusters, work-family-education balance, retirement and career wind down, organizational support for informal caregivers.

Jan. 2015 – Present
Coordinator
Réseau infirmier, un partenaire de soins (RIUPS).
Faculty of Nursing, Université de Montréal.

Oct. 2007 – June 2014
Coordinator and administrative assistant
Research professional
Community-University Research Alliances on "Work-life Articulation Over the Lifecourse" (CURA- WAROL) *www.teluq.ca/aruc-gats*
Canada Research Chair on Socio-Organizational Challenges of the Knowledge Economy - TÉLUQ

Dec. 2006 - May 2007
Research professional
Ministère de l'emploi et de la solidarité sociale et la Commission des partenaires du marché du travail (CPMT).

Jan. 2002 – Sep. 2006
Research and planning assistant

Centre interdisciplinaire de recherche/développement sur l'éducation permanente (CIRDEP) at Université du Québec à Montréal (UQAM).

Jan. 2000 - May 2002

Research assistant

Centre interuniversitaire de recherche sur la science et la technologie (CIRST) at Université du Québec à Montréal (UQAM).

Study conducted for the *Groupe interuniversitaire de recherche en formation – emploi (GIRFE)*

Sep. 2000 - June 2001

Teaching assistant

Qualitative methodology workshop – Required 3rd year undergraduate course.

Université du Québec à Montréal (UQAM).

March 2000 - July 2000

Research assistant

Comité sectoriel de la main-d'œuvre des technologies de l'information et des communications TechnoCompétences.

Publications Last 3 Years:

Lefebvre et al. (2015). Une innovation pour un meilleur continuum de soin en oncologie: développement et implantation d'un portail web d'échange des savoirs. *Revue francophone internationale de recherche infirmière*http://dx.doi.org/10.1016/j.refiri.2015.10.003

Diane-Gabrielle Tremblay, Maryse Larivière and Donna Lero (2013). *La conciliation emploi-famille-soins : quel soutien des entreprises pour les proches aidants?* Note de recherche no. 2013-07 de l'ARUC sur la gestion des âges et des temps sociaux.

Tremblay, Diane-Gabrielle and Larivière, Maryse (2013). *La conciliation emploi-famille-soins : quel soutien des entreprises pour les personnes qui doivent soutenir des parents vieillissants ou un conjoint malade ?* Note de recherche no 2013-1 de l'ARUC sur la gestion des âges et des temps sociaux.

Larivière, M., D-G, Tremblay. (2013). *Temps professionnels, temps prescrits, temps familiaux et personnels chez les périsoignants et organisateurs communautaires*. Note de recherche no 2013-6 de l'ARUC sur la gestion des âges et des temps sociaux.

Larivière, M., D-G, Tremblay, and E. Mascova. (2013). *Défis professionnels et personnels, action syndicale et paradoxes. La conciliation emploi famille chez les conseillers syndicaux et employés de bureau dans une*

organisation syndicale. Note de recherche No 2013-02 de l'ARUC sur la gestion des âges et des temps sociaux.
Tremblay, D-G. and M. Larivière. (2013). *La conciliation emploi-famille-soins : quel soutien des entreprises pour les proches aidants?* ARUC/GATS. Note de recherche no 2013-05 de l'ARUC sur la gestion des âges et des temps sociaux.

Michelle Proulx
Université de Montréal

Education: PhD, Public Health, Université de Montréal

Address: 801 Gameroff # 5
Lachine (Québec), Canada

Research and Professional Experience:

Michelle Proulx has been self-employed since fall 2008 and has worked on various freelance projects in nursing, education sciences, and for rehabilitation centers in Greater Montreal (qualitative analyses, drafting reports and grant applications, scientific articles). Since 2013, she has worked on several mandates for the team working on the "Developing a clinical innovation in oncology: toward an improved continuum of care and services" project, focusing on research activities (analyses, summaries) and knowledge transfer, including moderating the activities of the Forum for Knowledge Exchange (FKE). She also collaborates on knowledge transfer projects for the Centre for Interdisciplinary Research in Rehabilitation of Greater Montreal (CRIR) (consulting and preparing situation reports on knowledge acquisition in rehabilitation centers in Greater Montreal, supporting teams in the development of knowledge transfer projects in clinical settings) and works in the field of health promotion as part of the Interdisciplinary Research Group on Adaptation and Resilience (GIRR) (e.g., use of public places by and social participation of elderly people following a traumatic brain injury).

Conferences, Fieldwork and Research projects:

"Expérience de fréquentation des lieux publics par des personnes âgées avec TCC en présence d'un accompagnateur-citoyen : projet pilote." Analyses and summaries for the preparation of a scientific article. Interdisciplinary Research Group on Adaptation and Resilience (GIRR). 2014.

"Implementation Process and Impact of SAGEs (Specialists in the Application and Generalization of Expertise) as Change Agents in a Rehabilitation Center: The Clinician's Perspective." Quebec Rehabilitation Research Network (FRQS). 2014.

"L'expérience de la parentalité suite à un traumatisme cranio-cérébral d'un parent et les impacts chez les enfants: État de connaissances." Interdisciplinary Research Group on Adaptation and Resilience (GIRR). 2014.

"Supporting people with TBI in their use of public spaces: Identifying facilitating factors and obstacles." Faculty of Nursing. Université de Montréal. 2014.

Creating a databank using Zotero software for the Quebec Nursing Intervention Research Network (RRISIQ). Faculty of Nursing. Université de Montréal. 2013-2015.

"Évaluation qualitative sur la situation en matière d'appropriation des connaissances dans les établissements de réadaptation de la grande région de Montréal et intérêt pour une vision, des moyens et actions interétablissements." Brief on management and knowledge exchange for the RUIS of Université de Montréal and McGill University. 2013.

"Projet APIC : Un accompagnement citoyen personnalisé en soutien à l'intégration communautaire des personnes ayant subi un traumatisme cranio-cérébral: Vers la résilience?" Research report, Faculty of Nursing. Université de Montréal. 2013.

"Un service d'aide offert par des enseignants experts : la réussite de l'étudiant au cœur de leurs préoccupations." Analyses and drafting for the preparation of a scientific article. Faculty of Medicine and Health Sciences, Université de Sherbrooke. 2013.

"RESPECT: A Website to Deal with Mistreatment." Analyses and drafting for the preparation of a scientific article. Faculty of Medicine and Health Sciences, Université de Sherbrooke. 2013.

"Achieving a Maximum Level of Vaccination for Medical Students: A Rigorous Ethical and Legal Framework Procedure." Analyses and drafting for the preparation of a scientific article. Faculty of Medicine and Health Sciences, Université de Sherbrooke. 2013.

"Développement professionnel des enseignants: regard sur le projet d'un directeur adjoint de contre." Analyses and drafting for the preparation of a scientific article. Department of Education Management, Université de Sherbrooke. 2013

"La recherche-action-formation: rencontres et tensions." Analyses and drafting for the preparation of a scientific article. Department of Education Management, Université de Sherbrooke. 2013.

Professional Appointments:

Self-employed and freelance worker in the fields of health and education.

Grants:

Canadian Institutes of Health Research (CIHR) / Hypertension Canada, $38,000 (2003-2005).

ANÉIS Program (strategic training program in the analysis and assessment of health interventions, in partnership with CIHR/FRSQ), Université de Montréal, $5,000 (2004).

Faculty of Graduate Studies, Université de Montréal, $4,000 (2000); $2,500 (2001).

Health Promotion Research Centre, $10,500 (1993).

Publications:

Lefebvre, H., Brault, I., Levert, M.-J., Roy, O., Proulx, M., Alarie, S., Larivière, S. In collaboration Beauchamps, J., Carignan, M., Compagna, L., Déry, J., Leboeur, I., Tremblay, C. (2015). Le développement d'une innovation clinique en oncologie: pour un meilleur continuum de soins et de services pour les patients atteints de cancer. *Revue francophone internationale de recherche infirmière.* http://www.em-premium.com/article/1017666

Levert, M.-J., Lefebvre, H, Gélinas, I, McKerall, M, Roy, O, Proulx, M. (2015) "Expérience de fréquentation des lieux publics par des personnes âgées avec TCC en présence d'un accompagnateur-citoyen: projet pilote." *Revue canadienne du vieillissement* (accepted for publication).

Marie-Josée Levert

Assistant Professor, Faculty of Nursing, Université de Montréal Researcher, Centre for Interdisciplinary Research in Rehabilitation of Greater Montreal Co-Director, Interdisciplinary Research Group on Resilience and Community Rehabilitation

Education: Post-doctoral (Rehabilitation), McGill University; PhD (Psychology), Université du Québec à Montréal

Research and Professional Experience:

Dr. Levert's research focuses on developing innovative community interventions to support resilience of individuals living with various chronic diseases. She has held numerous grants as principal investigator and co-investigator, including grants from the CIHR and the FRQ-S. She has presented at many conferences, both nationally and internationally, and has co-authored over 60 peer-reviewed articles, monographs and book chapters.

Professional Appointments:

Assistant Professor, Faculty of Nursing, Université de Montréal

Honors:

People's Choice Poster Award: Brault, I., Lefebvre, H., **Levert, M-J,** Roy, O. (2015). A Web Knowledge Exchange Portal for a Better Discharge Planning in Oncology. International Conference on Cancer Nursing (ICCN), July 8-12, 2015. Vancouver, British Columbia, Canada.

Publications Last 3 Years:

Peer reviewed articles

Levert, M.-J., Gélinas, I., Lefebvre, H., McKerall, M., Roy, O., Proulx, M. (accepted). Expérience de fréquentation des lieux publics par des personnes âgées ayant subi un TCC en présence d'un accompagnateur-citoyen : projet pilote [Experience of public places frequentation by older adults who sustained a traumatic brain injury: a pilot project]. *Canadian Journal on Aging.*

Levert, M.-J., Lefebvre, H. (2014). L'accompagnement citoyen : une solution efficace pour soutenir le continuum de soins en post-réadaptation. *Le point en administration des services de santé, 9(4)*, 42-53.

Lefebvre, H., Levert, M.-J., LeDorze, G., Croteau, C., Gélinas, I., Therriault, P.-Y., Michallet, B., Samuelson, J. (2013). Un accompagnement citoyen personnalisé en soutien à l'intégration communautaire des personnes ayant subi un traumatisme craniocérébral : vers la résilience? *Recherche en soins infirmiers, 115*, 105-121.

Therriault, P.Y., Lefebvre, H, Guindon, A., Levert, M.-J., Briand, C. (2013). Accompagnement citoyen personnalisé en intégration communautaire : un défi pour la santé mentale ? *Santé mentale au Québec, 38*(1), 165-188.

Lefebvre H., Levert M.-J. (2012). Services vs. perceived needs: A status report in Québec and France on persons with moderate or severe traumatic brain injuries. *Journal of Trauma Nursing*, *19*(4), 197-207.

Lefebvre, H., Levert, M.-J. (2012). The close relatives of people who have had a traumatic brain injury and their special needs. *Brain Injury*, *26(9)*, 1084-1097.

Odette Roy
Associate Professor, Faculty of Nursing, Université de Montréal

Education: MSc. (Nursing), MAP (Masters in Public Administration), Ph.D.

Research and Professional Experience:

Lefebvre, H., Brault, I., and Roy, O. - Developing a clinical innovation in oncology: toward an improved continuum of care and services – Canadian Institutes of Health Research (CIHR) (PHSI) – Application submitted and approved – $660,000 grant - July 2013.

Lefebvre, H., Roy, O., Brault, I. et al., - Developing a clinical innovation in oncology: toward an improved continuum of care and services – *Fonds de recherche du Québec en santé* (FRQS) – Application accepted – $80,000 grant - January 2013.

Savignac, P., Roy, O., Foulem, J. – Exploration de l'efficacité d'une approche pédagogique en 4 temps visant le développement de capacités éthiques : prendre conscience de la situation, l'expression du souci éthique et l'écoute du récit – Funding from the HMR Foundation – $12,500 - May 2014.

Tremblay, N., Roy, O., et al. – Implantation d'un outil de la détresse psychologique atteinte de cancer : portrait de la clientèle et perception de la clientèle ainsi que des infirmières - Funding from the HMR Foundation ($75,000) and the Canadian Association of Nurses in Oncology ($5,000) - June 2014.

Malo., D., Levert, M.-J., Gélinas, I., Lefebvre, H., Mc Kerral, M., Roy, O. – Réseau TCCL - Préserver l'autonomie fonctionnelle des personnes âgées ayant subi un traumatisme craniocérébral léger (TCCL) : adaptation, implantation et évaluation de l'intervention d'accompagnement personnalisé d'intégration communautaire (APIC) - *Fonds de recherche du Québec en santé* (FRQS), Accepted – 2012 -

$149,728 grant.

Roy, O., Senneville, J. – Moyens pour prévenir les conséquences de l'immobilisation prolongée lors d'une hospitalisation – *Comité des usagers de l'Hôpital Maisonneuve-Rosemont* – June 2013 – $13,500 grant.

Professional Appointments:

Board of directors

Secrétariat international des infirmières et infirmiers de l'Espace Francophone (SIDIIEF) – Vice-president of the Board – *Consortium partenaire de l'Est* – 2014-2016.

Secrétariat international des infirmières et infirmiers de l'Espace Francophone (SIDIIEF) – Member of the Board – *Consortium partenaire de l'Est* - 2009-2014.

Scientific Committee – International symposium of the GRIISIQ (Quebec Nursing Intervention Research Network) – 2009-2012.

Board of directors of the FERASI Centre (Training and expertise in nursing administration research), Université de Montréal – Faculty of Sciences – 2008-2012.

Board of directors of the *Centre d'innovation en formation infirmière* (CIFI) – Université de Montréal – Faculty of Nursing – 2008-2012.

Scientific committees or committees of experts

Ordre des infirmières et infirmiers du Québec (OIIQ), Member of the committee – National Bank Innovation Award – 2012- 2015.

Ordre des infirmières et infirmiers du Québec (OIIQ), Member of the committee – Heritage Fund – 2011-2015.

FRESIQ, Member of the scientific committee. January 2001 to November 2005.

Professional committees

Ordre des infirmières et infirmiers du Québec (OIIQ). Committee of Heritage Fund Grants – (Clinical practice representative) – 2011-2013.

Ordre régional des infirmières et infirmiers de Montréal-Laval – Awards Committee – 2006-Present.

Ordre des infirmières et infirmiers du Québec (OIIQ). Professional Review Board – 2005-Present.

Honors:

Merit Insignia, *Ordre des infirmières et infirmiers du Québec* (OIIQ) - 2015;

Merit Award, Quebec Interprofessional Council (CIQ) - 2015;

Award of Recognition, *Secrétariat des infirmières et infirmiers de l'Espace Francophone* (SIDIIEF) – 2009;

Florence Award, *Ordre des infirmières et infirmiers du Québec* (OIIQ) - 2004.

Publications Last 3 Years:

Lefebre, H., Brault, I., Levert, M.-J., Roy, O., Proulx, M., Alarie, S., Larivière, M. "Une innovation pour un meilleur continuum de soin en oncologie: développement et implantation d'un portail web d'échange des savoirs." Revue francophone international de recherche infirmière.

Roy, O., Gagnon, R. La contention physique. In Voyer, P., *Soins aux personnes âgées.* Edition du renouveau pédagogique inc. 2013.

Lefebvre, H., Roy, O., Sahtali, S., Rothan-Tondeur, M. "*Lien entre recherche et pratique: échange des savoirs, utilisation des données probantes et renouvellement des pratiques." Revue SOINS,* 2014.

Roy, O., *Plaidoyer en faveur de la santé de la population: vers une pleine reconnaissance de la contribution infirmière* – SIDIIEF - Libreville (Gabon) – 2013.

Roy, O. *Centre d'excellence en soins infirmiers: 10 ans de leadership et d'innovation.* Résumé dans Colloque des CII-CIR 2013 – Culture de qualité et gestion des risques – Montréal, May 1, 2013.

Roy, O., *Le Centre d'excellence en soins infirmiers : structure innovante au service d'une culture de recherche en soins infirmiers* - Première journée francophone de la recherche en soins – Angers (France) – April 11-12, 2013.

Sophie Alarie
Université de Montréal

Education: M.Sc. Criminology
Université de Montréal

Address: Faculty of Nursing
Université de Montréal
CP. 6128 Succ. Centre-Ville
Montréal, Quebec Canada H3T 3J7

Research and Professional Experience:

Sophie Alarie has worked in the field of research since 1993 and has been a research professional at the Université de Montréal since 2004. Since 2003, she has coordinated the evaluation of different programs and from 2009 to 2013 she was the assistant to the director of the International Forum of Public Universities. Her research focus includes Program evaluation, evaluative research, drug/alcohol-related crimes, rehabilitation Interventions of criminalized or not drug/alcohol users, homelessness, the patient-as-partner in research, and harm Reduction.

Professional Appointments:

Program Evaluation, Drug/Alcohol-related crimes, Rehabilitation Interventions of Criminalized or Not Drug/Alcohol users, Homelessness, Harm Reduction.

Publications Last 3 Years:

Lefebvre, H., Brault, I., Levert M-.J., Roy O., Proulx, M., Alarie S., Larivière, M. (2015). Une innovation pour un meilleur continuum de soin en oncologie: développement et implantation d'un portail web d'échange des savoirs. *Revue francophone internationale de recherche infirmière*, 1-10. Doi: 10.1016/j.refiri.2015.10.003

Isabelle Brault
Université de Montréal

Education: PhD

Address:

Faculty of Nursing
Pavillon Marguerite-d'Youville
C.P. 6128 succ. Centre-ville
Montréal (Quebec) H3C 3J7

Research and Professional Experience:

Isabelle Brault, RN, PhD, is an Assistant Professor at the Faculty of Nursing, Université de Montréal, and a researcher at the Quebec Nursing Intervention Research Network and the *Institut de recherche en santé publique* of Université de Montréal. She is the leader of interfaculty interprofessional education (IPE) courses of Collaboration in health care at the Faculty of Nursing of Université de Montréal. Since June 2014, she has been vice president of the Interfaculty Operational Committee for Interprofessional Education on Collaboration and Partnership in Care. Her research focus includes evaluative research on interprofessional education on collaboration and partnership in care, nursing administration, clinical governance and knowledge transfer.

Professional Appointments:

Assistant Professor, Faculty of Nursing, Université de Montréal

Honors:

People's Choice Poster Award: Brault, I., Lefebvre, H., Levert, M-J, Roy, O. (2015). A Web Knowledge Exchange Portal for Better Discharge Planning in Oncology. International Conference on Cancer Nursing (ICCN), July 8-12, 2015. Vancouver, British Columbia, Canada.

Publications Last 3 Years:

Peer reviewed articles

Lefebvre, H., Brault, I., Levert, M-J., Roy, O., Proulx, M., Alaire, S., Larrivière (2015). Une innovation pour un meilleur continuum de soin en oncologie: développement et implantation d'un portail web d'échange des savoirs. Revue francophone internationale de recherche infirmière. http://dx.doi.org/10.1016/j.refiri.2015.10.003

Brault, I, Denis, J-L, Sullivan, T. (2015). Using Clinical Governance Levers to Support Change in a Cancer Care Reform. Journal of Health Organization and Management 2015; 29 (4) 482-97. doi: 10.1108/JHOM-02-2015-0025.

Brault, I, Therriault, PY, Lebel P, St-Denis, L. (2015) Implementation of interprofessional learning activities in a professional disciplinary practicum: the emerging role of technology. Journal of Interprofessional Care. 2015 May 8: 1-6.

Contandriopoulos, D, Brousselle, A, Dubois, C-A, Perroux M, Beaulieu M-D, Brault I, Kilpatrick K, D'Amour D and Sansgter-Gormley E (2015). A process-based framework to guide nurse practitioners integration into primary healthcare teams: results from a logic analysis. BMC Health Services Research, 15:78

Brault I, Kilpatrick K, D'Amour D, Contandriopoulos D, Chouinard, V, Dubois C-A, Perroux M, Beaulieu M-D (2014). Role Clarification Processes for Better Integration of Nurse Practitioners into Primary Healthcare Teams: A Multiple-Case Study. Nursing Research and Practice. Volume 2014 (2014), Article ID 170514, 9 pages. http://dx.doi.org/10.1155/2014/170514

Poisson, C., Alderson M., Caux, C., Brault, I. La détresse morale vécue par les infirmières: état des connaissances. Recherche en soins infirmiers, June 2014; 117: 65-74.

Vanier, M-C; Therriault, P-Y; Lebel, P; Nolin, F; Lefebvre, H; Brault, I; Drouin, E; Fernandez, N. (2013). Innovating in Teaching Collaborative Practice with a Large Student Cohort at Université de Montréal. Journal of Allied Health, Volume 42, Number 4, Winter 2013, pp. 97E-106E(10).

Dubois, C-A., Pomey, M-P., Girard, F., Brault, I. (2013). Conceptualizing performance of nursing care as a prerequisite for better measurement: a systematic and interpretive review. BMC Nursing, 12 (1).

Book Chapter

Lefebvre, H., Brault, I, Roy, O. (2015). *Une innovation clinique pour accompagner le changement dans les pratiques infirmières* dans Accompagner le changement dans le champ de la santé. De Boeck-Estem, éditeurs. Chapter 3; pp. 39-50.

In: Knowledge Creation and Transfer
Editor: Edwin Matos
ISBN: 978-1-63485-656-0

Chapter 4

THE KNOWLEDGE TRANSFER TO SUBSIDIARIES: THE IMPORTANCE OF LOCAL INTERNAL STAKEHOLDERS

José Moleiro Martins*
Department of Management, Lisbon Accounting and Business School
Lisboa, Portugal

ABSTRACT

The purpose of this chapter is to better understand the role of internal stakeholders in a developing country to offer potential insights into the cross-border transfer of knowledge. The study involves case studies of four companies that invest in Mozambique in business areas which are presently of interest. The results obtained reveal insights that linked successful knowledge transfer to overseas with a set of operational factors: different perception of local internal stakeholders of what is happening in organizational context; affirmation of the working philosophy of the parent company underlying the organizational practices transferred; operational relevance of the transferred knowledge; and adaptation of the practices to the local context. The case study method does not allow the generalization of the results, but it permits the study of a combination of problems of the phenomenon that contribute to its better understanding. This chapter contributes to multinational corporations to know how to transfer knowledge to their overseas subsidiaries that

* Corresponding author: Email: zdmmartins@gmail.com.

operate in the industrial sectors, taking into consideration the particular characteristics of the market and the absorptive capacity existing in lesser developed countries.

Keywords: knowledge transfer, culture, internal stakeholders, absorptive capacity

INTRODUCTION

One of the most fundamental questions to management researchers remains that of why firms are different, and why certain firms achieve and maintain competitive advantage. Competitive advantage is built on a unique bundle of assets that is difficult to imitate. Its sustainability depends on the continuous development of two key resources, one is culture and one knowledge (Prahalad and Hamel, 1990). Therefore, knowledge can be considered the most important strategic resource, and the ability to acquire, integrate, store, share, and apply it the most important capability for building and sustaining competitive advantage (Grant, 1996; Kogut and Zander, 1992).

Previous studies have highlighted the organizational, structural and relational aspects in international knowledge transfer (Björkman et al., 2004; Jensen and Szulanski, 2004), but did not consider the influence of local internal stakeholders on the acquisition, sharing, and applying knowledge channeled to the local market. A stakeholder in an organization is any group or individual who can affect or is affected by the achievement of the organization´s objectives (Freeman, 1984). Primary stakeholders include value chain participants and internal stakeholders (Donaldson and Preston, 1995). Internal stakeholders include management and non-management employees (Freeman, 1984).

The process of transfer occur in a relational context in terms of the past relationship between the parent company and the recipient unit, as perceived by the "key players" at the recipient unit involved in a particular transfer (Kostova, 1999). An initial necessary requirement for employees to become influential stakeholders is that the knowledge they possess is a critical asset to the firm (Kochan and Rubinstein, 2000).

This chapter has two aims in knowledge transfer to overseas: first, to help the executives at the transnational level to understand what influence the learning actions of the internal stakeholders registered in the subsidiary; second, to empirically evaluate which best practice are used locally to improve the absorptive capacity of the internal stakeholders and the performance of the subsidiary.

In the context of cross-border transfer of organizational learning, knowledge-oriented perspectives focus on knowledge transfer from the headquarters to the foreign subsidiaries and its use as a tool for generating further, context-specific, knowledge to address local problems and challenges (Gupta and Govindarajan, 2000).

The next section of the chapter presents the theoretical background for the study. The research methodology of the present study is explained followed by a presentation of the findings with regard to overcoming the challenges posed by local environments in order to supply the subsidiaries with the appropriate abilities for developing their economic activity. Finally, the chapter concludes with implications for practitioners which intend to improve the competencies of the local employees and consequently the stock of knowledge in the Mozambican subsidiaries.

Cross Cultural Knowledge Transfer

Today, knowledge is shared between geographically dispersed individuals and organizations, and cross cultural and national boundaries. The content of such knowledge may be technical, procedural or strategic (Child and Rodrigues, 1996), may be explicit or tacit (Nonaka and Takeuchi, 1995; Polanyi, 1966; Spender, 1996), and it may be encoded, embrained, encultured, embedded in various repositories (Blackler, 1995).

Organizational learning is influenced by various contexts, such as culture and history (Edmondson, 2002; Tyre and Von Hippel, 1997). The contexts constitute a complex system of knowledge, routines and other enabling conditions that drives the interpersonal and organizational dynamics within and between communities of practice (Brown and Duguid, 1991; Gherardi and Nicolini, 2002). A community of practice is a work-related group of individuals who share common interests or problems, and learn from each other through on-going interactions (Lave and Wenger, 1991). Knowledge sharing is embedded in broader organizational networks such as communities of practice. The ties among individuals within social networks can facilitate knowledge transfer and enhance the quality of information received (Cross and Cummings, 2004; Reagans and McEvily, 2003). Knowledge transfer involves both the sharing of knowledge by the knowledge source and acquisition and application of knowledge by the recipient (Wang and Noe, 2010). Knowledge transfer requires a co-operative and collaborative culture between source and recipient; it also facilitates knowledge transfer between developed and developing countries. The cultural differences between the parties (source, recipient) create extra problems because they make it difficult for managers to work together effectively and develop common values. For this reason, both organizational and national cultures are often referred to as a major obstacle, which has negative influences on all facets of cooperation encountered in inter-firm knowledge transfer (Mowery et al., 1996; Simonin, 1999).

Local employees base their trust in the organization on whether management is perceived to be benevolent (Pirson and Malhotra, 2011). Zaheer et al. (1998) argue that the origin of trust is always grounded in an individual perspective, even if individuals belonging to a certain group may share a similar orientation. We define trust as the psychological willingness of a party to be vulnerable to the actions of another party (individual or organizational) based on positive expectations regarding the other party's motivation and/or behavior (Ferrell, 2004; Mayer et al., 1995). Stakeholder trust in organizations entails the willingness of individuals to accept vulnerability to the actions of an organization based on positive expectations. Internal stakeholders, such as employees and investors, may care relatively more about managerial competence, such as decision-making ability and strategic vision (Pirson and Malhotra, 2011). Managerial competence is implicated in the organization ability to make strategic decisions and manage stakeholder relationships locally.

Knowledge is transferred via channels and activities such as training and education, business networking, development of new business opportunities and exchange of ideas. Knowledge transfer at the intra and transnational organizational level views an organization as an entity within which knowledge is created and shared. Therefore organizational practices, human resources abilities, structure, and processes affected by local environment, and play a critical role in articulating and amplifying knowledge developed by individuals (Hong et al., 2006). Individual team members contribute their distinct knowledge through their interactions with each other (Smith and Tushman, 2005), they are likely to be constrained in their individual capacity to attend to multiple stakeholders (Crilly and Sloan, 2014).

The inter-organizational trust is a critical component for knowledge acquisition as it encourages the knowledge transfer to assist the knowledge acquirer to understand the knowledge it is teaching. This promotes behaviors such as open communication and the willingness to share information (Norman, 2004). Kim (1998) argues that a critical component developing and maintaining individual absorptive capacity is closely related to the inspiration of intent to learn. Cohen and Levinthal (1990) contend that an organization's absorptive capacity is dependent upon the learning capabilities of its individual members. Based on their explanation, the fundamental premise of the notion of absorptive capacity is that firms need prior related knowledge to evaluate and utilize outside information.

Absorptive Capacity of Subsidiaries

Cohen and Levinthal (1990) suggest that the absorptive capacity is a prerequisite for organizational learning in the sense that it refers to internal resources that strengthen a firm's ability to understand, assimilate new knowledge and apply it to commercial ends. Absorptive capacity has received a lot of attention from the research community as the most significant determinant of organizational learning. Kim (1998) defined it as the combination of prior knowledge base and intensity of effort. It has also been defined by Zahra and George (2002) as a set of organizational routines and processes, by which firms acquire, assimilate, transform and exploit knowledge to produce a dynamic organizational capability. Thus, the greater the absorptive capacity of the subsidiary, the more knowledge it can acquire from headquarters.

The inability of knowledge receivers to recognize the value of new information, assimilate it, and apply it to commercial ends is one of the most often referred impediments to knowledge transfer. Both employees' ability and motivation need to be present in order optimally to facilitate the absorption of knowledge from other parts of the multinational corporations (MNCs) (Minbaeva et al., 2003).

The low absorptive capacity of Mozambican subsidiaries makes it necessary to make efforts with respect to the transfer of knowledge from the MNC headquarters in order to improve the competencies of the local employees.

The purpose of the following is to better understand organizational learning in subsidiaries, and the role of internal stakeholders in a developing country to offer potential insights into the cross-border transfer of knowledge

This study takes into account the challenges posed by local internal stakeholders (i.e., workers) as important agents of organizational learning in overseas. This study explored the learning capability applied of local employees and, consequently, the absorptive capacity of the Mozambican subsidiaries as a competitive source on local market.

METHODOLOGY

The research is multiple-case, which permits a logic replication (Yin, 1994). The study involved case studies of four Portuguese companies that invest in Mozambique in business areas which are presently of interest in industrial sector.

I gathered information on the perspectives of internal stakeholders (i.e., middle and top levels of the management hierarchy).

I collected data through semistructured interviews with individual respondents. At both the top and middle management, there was a mix of operational, strategy, commercial and engineering informants. All interviews were conducted at the research site, taped and transcribed. The sixteen interview participants all had deep knowledge of the organization's structure, culture, and best practices.

I used one interview guide to conduct the semistructured interviews. I analyzed the data by first building individual case studies and then comparing across cases to construct a conceptual framework (Eisenhardt, 1989). Once the individual case studies were complete, I used a cross-case analysis, relying on methods suggested by Miles and Huberman (1984), to develop the conceptual insights. This iterative process led to the insights that follow.

FINDINGS

The major results from the study are concerning the two organizational dimensions that permit to answer the difficulties that emerge in the process of knowledge transfer in order to improve the subsidiary's performance. Two research questions are answered.

Question 1: What is it that influences the learning actions of local internal stakeholders?

The results revealed that the stock of knowledge on the part of the recipient is very low, and exists a common trace with respect of the knowledge flows to subsidiaries as indicated by a headquarters executive: "we possess human resources who know the African market and have availability to transmit knowledge and create a good information flow between headquarters and the subsidiaries" (case 2).

The evidence of the results suggests that "must be avoided to present the transfer of knowledge as a declared imposition, so that the employees of the subsidiary may get involved in this process and feel the necessity of this knowledge" (case 4). The technological abilities are transmitted by the sharing of technical knowledge on the part of key personnel who teaches by doing in the field of practices. The interviewed managers agree that "the organizational culture and philosophy of the parent MNC can be transmitted during the on-the-job training" (case 1). This procedure facilitates the "individual learning via the participation in actions in which experiences are demonstrated and shared and which are accompanied by the exchange of opinions among the participants" (case 3). "The errors are tolerated in order to create a good relationship between people and create a common language that 'qualifies' people for the acquisition and assimilation of the new knowledge" (case 4). The results are consistent with the study of Kochan and Rubinstein (2000), since some collective means of mobilizing knowledge is likely to be needed for the larger number of workers whose potential value added is more likely to come from their joint efforts. Contributing value assets is only part of the requirement for gaining legitimacy as a stakeholder.

"At the basis of an effective relation of knowledge transfer is always a relationship of trust. Precisely because of this criterion, it will always be easier to achieve an effective development of mechanisms for the transfer of knowledge to a country that possesses a culture similar to ours, as is the case with Mozambique" (case 1). As explained another manager, "the culture and values of the parent company have to be shared, respecting the local cultural values" (case 2). As referred by Jones et al. (2007), stakeholder culture is a central facet of organizational culture, can provide managers with guidance regarding how this tension should be resolved.

At the beginning of the local operative activities, the managers "tried to apply the domestic business model, but did not succeed in rendering the installed production capacity lucrative" (case 3). This study shows that sometimes the results of the transfer of knowledge are not positive at the first attempt, which originates the introduction of alterations as compared to what was initially planned. From the point of view of the receiver "we may be seen as having certain aloofness with respect to the local culture, and this generates resistance in people and creates difficulties as far as the recognition of the importance of the transferred knowledge is concerned" (case 1). A local

Mozambican manager explained that "the individuals who do not possess a superior academic education have greater difficulties in saying that they don't know" (case 4).

The non-existence of qualified human resources leads to the formation of "technical barriers" (cases 1 and 3). In their interviews the managers stated that these obstacles are overcome by exposing the subsidiary's employees to the work environment of the headquarters in order for them to develop an effective control of the new knowledge that turns its application sustainable during a long period of time.

The knowledge transfer goes well if it is conducted adequately, i.e., without being openly imposed so that the people themselves may feel the need to learn (cases 1 and 4). On the other hand, if headquarters avoids transferring knowledge that is not valued by the subsidiary despite its requesting new knowledge (cases 2 and 3).

Basically, the intention is to make the transfer easier through the receptiveness, understanding and sharing of knowledge by means of training in a work environment. Thus the Mozambican managers can share experiences with a sense of belonging associated to the objectives of the subsidiary, with a view to achieving operative improvements by increasing the absorptive capacity of the subsidiary, as will be seen in the following.

Question 2: Which are the best practices used locally to improve the absorptive capacity of the internal stakeholders of the subsidiary?

The ability of the employees to acquire and absorb new knowledge depends on whether the transferred knowledge possesses an operational pattern that is sufficiently flexible to allow its adjustment to the organizational format of the subsidiary, in order to be well understood by the local technicians (cases 1 and 4). The results show that "the subsidiary's ability to learn is affected by the individuals' lack of interest in what is being transmitted to them" (case 1). As the expatriate managers interviewed said, the MNC's stock of knowledge inherent to its core business can be replicated in Mozambique, but with an inferior degree of development (case 2).

Our results are consistent with Jones et al. (2007), inasmuch as internal stakeholder culture represents a firm's collective reconciliation of these contradictory motives and, as such consists of its shared beliefs, values, and evolved practices regarding the solution of recurring stakeholders-related problems. Stakeholder culture is grounded in ethics and is based on a continuum of concerns for others that runs from self-regarding to other-regarding. "Mozambicans reveal a different work ethic in relation to job duties" (case 4). This "different perception of what happens affects their learning" (case 1) in so far as "they view lack of success as a personal handicap" (case 1). Mangers explains that this is the result of the low educational levels of the Mozambicans which become obvious in their withdrawal from the search for knowledge which raises difficulties to the transfer of knowledge by way of the resistance to change in the organizational culture of the subsidiary. In line with Berman et al. (1999), were adopted HR practices to lower turnover and absenteeism, improve productivity, and increase worker commitment and effort.

For a stakeholder firm to function successfully, employee discretionary efforts need to be mobilized, high levels of communication and coordination are needed across groups and functions, and conflicts need to be surfaced and resolved effectively. Mozambican workers are invited to visit productive installations that are technologically more advanced. In case 3 "the administration of intensive courses in the productive installations of the source is the main measure adopted to increase the abilities of the employees in the subsidiary". Thus, "they can accompany the execution of tasks and get an idea of the production process as a whole" (case 1), which "helps them to understand the advantages that result for the subsidiary from the successful realization of this transfer of knowledge" (case 4).

To be influential stakeholders, employees must add value by using their knowledge and discretionary efforts to enhance firm performance. The interviews suggested that "the acquisition of the technical skills increase the subsidiary's competitiveness through the gradual transfer of practices adjusted to the particular local context" (case 1), with the "creation of new routines in the productive process that improve its performance" (case 4). In cases 1 and 4, the absorptive capacity of external knowledge contributes to the improvement of the recipient's productivity through new functional processes that reduce the duration of certain tasks, and in case 3 with gains in the ability to respond to the demands of the market. As mentioned by a top manager "the retention of knowledge is shown by the business results" (case 3). The knowledge transfer "added value to the products and transmits the image of an institution that is contributing to the development of the Mozambican market" (case 4). The results reflect the two-way relationship between a firm and its internal stakeholders. If internal stakeholders can affect the achievement of a firm's objectives, it follows that the firm's decisions, and hence its performance, may be affected by the activities of its internal stakeholders.

CONCLUSION

This research provides information for practitioners who may be helping their organizations through a better understanding of the knowledge transfer to subsidiary companies on competition in an unstable business environment as the one in Mozambique.

This study shows that the employees (i.e., internal stakeholders) are the source of a company's success, and successful environmental policy planning requires their participation and motivation. Internal stakeholder culture is influenced by perceptions of organizational values and the typical practices and procedures that have ethical content or pertain to moral behavior locally.

It is fundamental to convince the local people of the usefulness of the knowledge to be transferred from the headquarters to the subsidiary. Hence the importance of a good interpersonal relationship among the participants on both sides. The process must be conducted in a way that the Mozambican workers feel to be an active part in the operations to be carried out and thus are motivated to acquire technical skills through the assimilation of knowledge during the execution of experiences in interaction with the transmitter.

This study points to some important practical lessons. The poor qualifications of the local people are causing a time-lag of perceptions as compared with the parameters of functioning of the MNE home base, influenced by cultural values concentrated on the family in detriment to the professional duties and, as a consequence, responsible for the low productivity of work. This prevents the reproduction of the headquarters' operational performance in the subsidiary, i.e., it does not advise in favour of the replication of the home business model on the Mozambican market.

REFERENCES

Berman, S., Andrew, W., Kotha, S. and Jones, T. (1999). Does stakeholder orientation matter? The relationship between stakeholder management models and firm financial performance. *Academy of Management Journal*, Vol. *42*, No. 5, pp. 488-506.

Blackler, F. (1995). Knowledge, knowledge work and organizations – an overview and interpretation. *Organization Studies*, Vol. *16*, No. 6, pp. 1021-1046.

Brown, J. S. and Duguid, P. (1991). Organizational learning and communities of practice – toward a unified view of working, learning, and innovation. *Organization Science*, Vol. 2, No. 1, pp. 40-57.

Child, J. and Rodrigues, S. (1996). The role of social identity in the international transfer of knowledge through joint ventures. In S. Clegg and G. Palmer (Eds.), *The Politics of Management Knowledge*, (pp. 46-68). London, Sage.

Cohen, W. M. and Levinthal, D. A. (1990). Absorptive capacity – a new perspective on learning and innovation. *Administrative Science Quarterly*, Vol. *35*, No. 1, pp. 128-152.

Crilly, D. and Sloan, P. (2014). Autonomy or control? Organizational architecture and corporate attention to stakeholders. *Organization Science*, Vol. *25*, No. 2, pp. 339-355.

Cross, R. and Cummings, J. N. (2004). Tie and network correlates of individual performance in knowledge-intensive work. *Academy of Management Journal*, Vol. *47*, No. 6, pp. 928-937.

Donaldson, T. and Preston, L. (1995). The stakeholder theory of the corporation: Concepts, evidence, and implications. *Academy of Management Review*, Vol. *20*, No. 1, pp. 65-91.

Edmondson, A. (2002). The local and variegated nature of learning in organizations – a group-level perspective. *Organization Science*, Vol. *13*, No. 2, pp. 128-146.

Eisenhardt, K. M. (1989). Building theories from case study research. *Academy of Management Review*, Vol. *14*, No. 4, pp. 532-550.

Ferrell, O. (2004). Business ethics and customer stakeholders. *Academy of Management Executive*, Vol. *18*, No. 2, pp. 126–129.

Freeman, R. (1984). *Strategic Management: A Stakeholder Approach*. Boston, MA, Pitman.

Gherardi, S. and Nicolini, D. (2002). Learning in a constellation of interconnected practices: Canon or Dissonance? *Journal of Management Studies*, Vol. *39*, No. 4, pp. 419-436.

Gupta, A. K. and Govindarajan, V. (2000). Knowledge flows within multinational corporations. *Strategic Management Journal*, Vol. *21*, No. 4, pp. 473-496.

Hong, J., Easterby-Smith, M. and Snell, R. (2006). Transferring organizational learning systems to Japanese subsidiaries in China. *Journal of Management Studies*, Vol. *43*, No. 5, pp. 1027-1058.

Jones, T., Felps, W. and Bigley, G. (2007). Ethical theory and stakeholder-related decisions: The role of stakeholder culture. *Academy of Management Review*, Vol. *32*, No. 1, pp. 137-155.

Kim, L. (1998). Crisis construction and organizational learning: Capability building in catching-up at Hyundai Motor. *Organization Science*, Vol. *9*, No. 4, pp. 506-521.

Kochan, T. and Rubinstein, S. (2000). Toward a stakeholder theory of the firm: The Saturn partnership. *Organization Science*, Vol. *11*, No. 4, pp. 367-386.

Kostova, T. (1999). Transnational transfer of strategic organizational practices: A contextual perspective. *Academy of Management Review*, Vol. *24*, No. 2, pp. 308-324.

Lave, J. and Wenger, E. (1991). *Situated Learning: Legitimate Peripheral Participation*. Cambridge, Cambridge, University Press.

Mayer, R., Davis, J. and Schoorman, F. (1995). An integrative model of organizational trust. *Academy of Management Review*, Vol. *20*, No. 3, pp. 709–734.

Miles, M. B. and Huberman, A. M. (1984). *Qualitative Data Analysis: A Sourcebook of New Methods*. Beverly Hills, Sage Publications.

Minbaeva, D., Pedersen, T., Björkman, I., Fey, C. F. and Park, H. J. (2003). MNC knowledge transfer, subsidiary absorptive capacity, and HRM. *Journal of International Business*, Vol. *34*, No. 6, pp. 586-599.

Mowery, D. C., Oxley, J. E. and Silverman, B. S. (1996). Strategic alliances and interfirm knowledge transfer. *Strategic Management Journal*, Vol. *17*, Winter Special Issue, pp. 77-91.

Nonaka, I. and Takeuchi, K. (1995). *The Knowledge Creating Company*. New York, Oxford University Press.

Norman, P. M. (2004). Knowledge acquisition, knowledge loss, and satisfaction in high technology alliances. *Journal of Business Research*, Vol. *57*, No. 6, pp. 610-619.

Pirson, M. and Malhotra, D. (2011). Foundations of organizational trust: What matters to different stakeholders? *Organization Science*, Vol. *22*, No. 1, pp. 1087-1104.

Polanyi, M. (1966). *The Tacit Dimension*. New York, Anchor Books.

Prahalad, C. K. and Hamel, G. (1990). The core competence of the corporation. *Harvard Business Review*, May-June, 79-91.

Reagans, R. and McEvily, B. (2003). Network structure and knowledge transfer: The effects of cohesion and range. *Administrative Science Quarterly*, Vol. *48*, No. 2, pp. 240-267.

Simonin, B. L. (1999). Ambiguity and the process of knowledge transfer in strategic alliances. *Strategic Management Journal*, Vol. *20*, No. 7, pp. 595-623.

Smith, W. and Tushman, M. (2005). Managing strategic contradictions: A top management model for managing innovation streams. *Organization Science*, Vol. *16*, No. 5, pp. 522-536.

Spender, J. C. (1996). Making knowledge the basis of a dynamic theory of the firm. *Strategic Management Journal*, Vol. *17*, Winter, pp. 45-62.

Tyre, M. J. and Von Hippel, E. (1997). The situated nature of adaptive learning in organizations. *Organization Science*, Vol. *8*, No. 1, pp. 71-83.

Wang, S. and Noe, R. (2010). Knowledge sharing: A review and directions for future research. *Human Resource Management Review*, Vol. *20*, No. 2, pp. 115-131.

Yin, R. K. (1994). *Case Study Research: Design and Methods*. London, Sage.

Zahra, S. A. and George, G. (2002). Absorptive capacity: A review, reconceptualization, and extension. *Academy of Management Review*, Vol. *27*, No. 2, pp. 185-203.

Zaheer, A., McEvily, B. and Perrone, V. (1998). Does trust matter? Exploring the effects of interorganizational and interpersonal trust on performance. *Organization Science*, Vol. *9*, No. 2, pp. 141–159.

Biographical Sketch

José Duarte Moleiro Martins
Lisbon Accounting and Business School

Education: PhD in Strategic Management from ISCTE-Lisbon University Institute.

Research and Professional Experience: Professor of Strategic Management and Entrepreneurship at Lisbon Accounting and Business School. He is also researcher at UNIDE-IUL research group in General Management Studies. His special current research interests are in foreign direct investment in Africa.

Publications Last 3 Years:

Martins, J. (2012), "Analytical dimensions of knowledge transfer to the subsidiaries", *European Business Review*, Vol. 24 Iss: 5, pp. 465-477.

Fernandes, M. and Martins, J. (2014), "Reduce risks, avoide failure", Innovation in SMEs – Guest editorial, *International Trade News – Journal of Economics*, 19-Feb-2014.

Martins, J. (2014), "Operating factors that help to improve performance in subsidiaries with low absorptive capacity", *Journal of East-West Business*, Vol. 20, Iss. 3, pp. 162-183.

Martins, J. and Fernandes, M. (2015), "Too small to innovate? Creating value with fewer resources", *Journal of Business Strategy*, Vol. 36, Iss. 2, pp. 25-33.

BIBLIOGRAPHY

Art, archaeology and science: an interdisciplinary approach to Chinese archaeological and artistic materials / Meili Yang. LCCN: 2015021461 Personal name: Yang, Meili. Edition: First published in 2016 in Great Britain by Sussex Academic Press. Published/Produced: Brighton: Sussex Academic Press, [2016] ISBN: 9781845197339 (hardcover: alkaline paper) LC classification: CC75 .Y36 2016 Summary: "Science has played a crucial role in the initial stages of westerners' collecting and studying ancient Chinese artworks and archaeological materials. The relevant scientific data were quickly generated using progressively sophisticated examination techniques. But the task of integrating scientific data with archaeological and art historical resources, and developing a workable interdisciplinary study method, has lagged behind scientific endeavor. Part One of this book explores the modes and functions of the scientific resources available and their integration into interdisciplinary study methods and models; pre-war researchers particularly emphasized that their studies of Chinese artistic/archaeological materials were scientific. Improvements in precision and advanced specialization of examining instruments and techniques resulted in substantively more detailed scientific data, particularly in the post-war period. This brought with it new research dimensions and increased knowledge of targeted samples, and also greater challenges to the integration of interdisciplinary study. In Part Two, Meili Yang establishes a feasible method of interdisciplinary study in terms of five case studies related to Chinese Song ceramics. Relevant data are provided and professionally explained, resulting in critical new information related to contemporary ceramic technologies. This technology information leads us to the recognition of the uniqueness of each single artwork and the artisan's individual intention - and critically, recognition of the close association with contemporary society and culture at the time of manufacture"-- Provided by publisher. Contents: Introduction -- Science, art and archaeology -- Scientific study of archaeological materials during the Chinese Song Dynasty: a Shen Kua case study -- The role of science in western Chinese jade collection and knowledge creation: a case study -- Modern western science in initial Chinese archaeology development -- Scientific study of Chinese archaeological materials in the pre-war period: Berthold Laufer's Han porcelain case study -- Qinghe and Song shards: rethinking Palmgren's scientific investigation in pre-war China -- Ceramic technology and art in Song society -- Ancient Chinese shards in the post-war West: science and art perspectives -- Technological variations and the artist's intention: a case study of Chinese southern Song Guan imperial ware -- Technology and art in society: the significance of "furnace transmutation" in Chinese Song Jun ware -- Sintering technology and art in Chinese Song Ding white ware -- Beyond cross-technology: a case study of Chinese Song imperial Ru in Jun glaze. Subjects: Archaeology--Methodology. Archaeology--Technological innovations. Archaeology and art--Case studies. Pottery, Chinese--Song-Yuan dynasties, 960-1368--Case studies. China--Antiquities--Case studies. Notes: Includes bibliographical references and index.

Beyond territory: dynamic geographies of knowledge creation, diffusion, and innovation / edited by Harald Bathelt, Maryann P. Feldman and Dieter F. Kogler. LCCN: 2010049284 Published/Created: New York: Routledge, 2011. Description: xvii, 291 p.: ill., map; 24 cm. ISBN: 9780415493277 (hardback) LC classification: HC79.T4 B493 2011 Related names: Bathelt, Harald. Feldman, Maryann P. Kogler, Dieter F., 1973- Summary: "The main purpose of the book is to discuss new trends in the dynamic geography of innovation and argue that in an era of increasing globalization, two trends seem quite dominant: rigid territorial models of innovation, and localized configurations of innovative activities. The book brings together scholars who are working on these topics. Rather than focusing on established concepts and theories, the book aims to question narrow explanations, rigid territorializations, and simplistic policy frameworks; it provides evidence that innovation, while not exclusively dependent on regional contexts, can be influenced by place-specific attributes. The study of innovation encompasses an increasingly wide and rich field of conceptual and empirical studies and debates, which span across disciplinary boundaries in the social sciences. This book identifies the key debates, new streams of inquiry and progress in research related to the transfer, circulation and generation of knowledge in a spatial perspective. The book is organized into three subheadings: place-specific aspects of specialization and diversity; evolutionary spatio-sectoral dynamics; and bridging the local and the global divide. It is the aggregate of all of these contributions that offers an exclusive insight into the dynamic geographies of knowledge creation, diffusion and innovation, well beyond the traditional territorial approach. This volume contains new empirical and conceptual work by an interdisciplinary group of leading scholars from areas such as economic geography, innovation studies, and political science"-- Provided by publisher. Subjects: Technological innovations. Knowledge management. Technology transfer. Globalization. Notes: Includes bibliographical references and index. Series: Regions and cities

Branded spaces: experience enactments and entanglements / Stephan Sonnenburg, Laura Baker (eds.). LCCN: 2013932217 Published/Produced: Wiesbaden: Springer VS, [2013] Description: 286 pages: illustrations, maps; 21 cm. Links: Contributor biographical information http://www.loc.gov/catdir/enhancements/fy1607/2013932217-b.html Publisher description http://www.loc.gov/catdir/enhancements/fy1607/2013932217-d.html Table of contents only http://www.loc.gov/catdir/enhancements/fy1607/2013932217-t.html ISBN: 9783658015602 3658015608 LC classification: HF5415.1255 .B7175 2013 Related names: Sonnenburg, Stephan, editor. Baker, Laura, editor. Balakrishnan, Melodena Stephens. 4D model of place brand management. Contents: Places and Possibilities -- The 4D Model of Place Brand Management / Melodena Stephens Balakrishnan, Greg Kerr -- Architecture and the Destination Image: Something Familiar, Something New, Something Virtual, Something True / Jan Specht -- Branding Views Marketing: Lessons to Learn for Destination Management / Prof. Dr. Werner Gronau PhD, Nicholas Adjouri -- The Possibility of a Social Imaginary: Public Housing as a Tool for City Branding / Michael Klein, Andreas Rumpfhuber -- Facts and Figures -- Corporate Branded Spaces: A Glance at Today's Diversity and Some Historic Origins / Prof. Dr. Louise Bielzer -- Brand Value in Real Estate: The Financial Contribution of Property Brands / Prof. Dr. Bernd Ankenbrand, Sven Mussler, Timo Mussler -- Naming Rights for Sports Stadiums: Sponsor versus Club -- Who Makes the Proper Brand? / Thomas Bezold -- Senses and Sensualities -- The Branded Experience: Decoding the Spatial Configuration of Flagship Stores / Garyfalia Palaiologou, Alan Penn -- Applying Theory of Branding Synergy on Consumer Experience and Spatial Design: A Case Study / Erica Liu -- Heimatklänge: The Conceptual Design of Branded Spaces by Means of Sonic Branding / Dr. Sonja Kastner -- Stories and Situations -- South Tyrol: Destinations Can Be Brands Too / Christoph Engl -- A Guggenheim in Every City? / Prof. Dr. Antti Vihinen -- AAALBANIA: Possibilities and Limitations of Nation Branding Using the Example of Albania / Jörg Dauscher, Frank Otto Dietrich, Ralf Schmidt-Bleeker -- Critiques and Consequences -- Branded Space; Branded Consumers: Spaces for Consumption and the Uncomfortable Consequences of

Complicit Communality / Steven Miles -- Cannibal Architecture Hates BANANAs: Post-Communist Rebranding of Historical Sites / Florin Mureşanu, Monica Mureşanu -- Syntax of Intervention in Historically Significant Public Open Urban Spaces / Stefanie Leontiadis PhD -- Branding as Enabling Knowledge Creation: The Role of Space and Cognition in Branding Processes / Markus F. Peschl, Thomas Fundneider. Subjects: Branding (Marketing) Branding (Marketing) Markenpolitik. Raum. Notes: Includes bibliographical references. Series: Management - culture - interpretation Management-Culture-Interpretation.

Connecting knowledge and performance in public services: from knowing to doing / edited by Kieran Walshe, Gill Harvey, Pauline Jas. LCCN: 2010021907 Published/Created: Cambridge; New York: Cambridge University Press, 2010. Description: xviii, 295 p.: ill.; 26 cm. Links: Cover image http://assets.cambridge.org/97805211/95461/cover/9780521195461.jpg Contributor biographical information http://www.loc.gov/catdir/enhancements/fy1010/2010021907-b.html Publisher description http://www.loc.gov/catdir/enhancements/fy1010/2010021907-d.html Table of contents only http://www.loc.gov/catdir/enhancements/fy1010/2010021907-t.html ISBN: 9780521195461 LC classification: HD4431 .C66 2010 Related names: Walshe, Kieran. Harvey, Gill, RGN. Jas, Pauline. Summary: "The performance of public services is now more closely scrutinised than ever before. Every teacher, doctor, social worker or probation officer knows that behind them stands a restless army of overseers, equipped with a panoply of league tables, star ratings, user opinion surveys, performance indicators and the like with which to judge them. This increased scrutiny and performance measurement has undoubtedly produced improved public services. Yet we still have a limited understanding about how this information can be best used to bring about improvements in performance. What goes on inside the 'black box' of public organisations to move from information to action, or from 'knowing' to 'doing'? This book tackles this important question by reviewing a wide range of performance mechanisms. It explores how information about performance can be translated into improvements in services and, conversely, why this does not always happen in practice"-- Provided by publisher. Contents: Machine generated contents note: List of figures; List of tables; List of boxes; Foreword Chris Hood; 1. Introduction: knowledge and performance - theory and practice Kieran Walshe, Gill Harvey and Pauline Jas; 2. Knowledge from inspection: external oversight and information to improve performance Steve Martin; 3. How is information used to improve public performance? Exploring the dynamics of performance information Steven Van de Walle and Wouter Van Dooren; 4. Citizens, users or consumers: the voice of the public and its influence on improving performance Ian Greener; 5. Competition and contestability: the place of markets in connecting information and performance improvement Carol Propper and Deborah Wilson; 6. The role of corporate governance and boards in organizational performance Chris Cornforth and Naomi Chambers; 7. Change at the top: connecting political and managerial transition with performance George Boyne, Oliver James, Peter John and Nicolai Petrovsky; 8. The role of leadership in knowledge creation and transfer for organizational learning and improvement Jean Hartley and Lyndsay Rashman; 9. Process improvement and lean thinking: using knowledge and information to improve performance Zoe Radnor; 10. Using evidence: how social research could be better used to improve public service performance Huw Davies, Sandra Nutley and Isabel Walter; 11. Absorptive capacity: how organizations assimilate and apply knowledge to improve performance Gill Harvey, Pauline Jas, Kieran Walshe and Chris Skelcher; 12. Knowing through doing: unleashing latent dynamic capabilities in the public sector Ann Casebeer, Trish Reay, James Dewald and Amy Pablo; 13. Conclusions: a puzzle, three pieces, many theories and a problem Colin Talbot; Index. Subjects: Municipal services--Management. Notes: Includes bibliographical references and index.

Cooperation, clusters, and knowledge transfer: universities and firms towards regional competitiveness / João J.M. Ferreira [and others], editors. LCCN: 2013933386 Published/Created: Berlin; New York: Springer, ©2013. Description: xx, 263 pages: illustrations; 25 cm. ISBN: 9783642331930 (hd. bd.) 3642331939 (hd. bd.) 3642331947 (electronic bk.) 9783642331947 (electronic bk.) Related names: Ferreira, João J. M. Summary: Cooperation and clusters have become the guiding paradigms for explaining and promoting regional competitiveness, but the cooperation process between firms and universities and the transfer of knowledge in guiding and nurturing regional competitiveness has received relatively little attention. This book strives to fill this gap in highlighting the connection between inter-firm cooperation in regional clusters, innovation and regional networks, and the role of universities in them . It goes beyond the traditional economic approach of clusters and includes 'soft factors' in the explanation of regional competitiveness, and connects the literature on clusters to the literature of learning and knowledge creation as sources of regional competitiveness. It aims to foster an international and interdisciplinary exchange of perspectives by presenting current developments, case studies, best practices as well as new integrated theoretical approaches and applications. Contents: The start-up location decision and regional determinants / Frank Lasch [and others] -- The role of science and technology parks in the generation of firm level social capital through university-firm relations: an empirical study in Spain / Juan J. Jimenez-Moreno [and others] -- Knowledge transfer in or through clusters: outline of a situated approach / Julie Hermans -- How does a researcher become an entrepreneur in the high-tech industrial cluster? A case study / Rongzhi Liu, Haiyan Zhang, and Zhi Yang -- The impact of inter-firm cooperation on performance: a two-region experience / João J.M. Ferreira, Mário L. Raposo, Cristina I. Fernandes -- Transition regions: green innovation and economic development / Philip Cooke -- Clusters, learning, and regional development: theory and cases / Roel Rutten and Dessy Irawati -- Sociocultural factors and female entrepreeurship in the innovative service sector in Catalonia: a qualitative analysis / Maria Noguera [and others] -- Academic entrepreneurship framework: the best practices of Bragança Polytechnic Institute / José Adriano [and others] -- Academic interactions with private, public and not-for-profit organisations: the known unknowns / Maria Abreu and Vadim Grinevich -- The role of academic spin-off founder's motivation in the Hungarian biotechnology sector / Katalin Erdős and Attila Varga -- Hirschman mobility, governance and loyalty in Europe's top research universities / Edward M. Bergman -- Action-based education in academic entrepreneurship: a new role of the student? / Lene Foss, Elin M. Oftedal, and Tatiana Iakovleva. Subjects: Technology transfer--Economic aspects. Small business--Technological innovations. Competition. Notes: Includes bibliographical references. Series: Advances in spatial science, 1430-9602 Advances in spatial science. 1430-9602

Creating collaborative advantage: innovation and knowledge creation in regional economies / edited by Hans Christian Garmann Johnsen and Richard Ennals. LCCN: 2011035068 Published/Created: Farnham; Burlington, VT: Gower, c2012. Description: xvi, 278 p.; 26 cm. ISBN: 9781409403333 (hbk.: alk. paper) 9781409403340 (ebook) LC classification: HD2963 .C74 2012 Related names: Johnsen, Hans Chr. Garmann, 1955- Ennals, J. R. (John Richard), 1951- Subjects: Cooperation. Strategic alliances (Business) Business networks. Public-private sector cooperation. Diffusion of innovations--Economic aspects. Research, Industrial. Regional economics. Notes: Includes bibliographical references (p. 253-274) and index.

Cross-Cutting Symposium: theme: "knowledge creation, incubation, and sharing for sustainable development". LCCN: 2014321403 Meeting name: Annual International Conference (10th: 2014: Moi University) Published/Produced: Eldoret, Kenya: Moi University Press, Moi University, 2014. Description: viii, 384 pages: illustrations; 21 cm ISBN: 9789966854766 LC classification: HD76 .A56 2014 Cover title: "Knowledge creation, incubation and sharing for sustainable development": symposium IV Spine title: 10th Annual International Conference: symposium IV Subjects: Communication in economic development--Kenya--Congresses. Notes: Includes bibliographical references.

Cultural diversity: issues, challenges and perspectives / Lydia B. Kerwin, editor. LCCN: 2009032127 Published/Created: New York: Nova Science Publishers, c2010. Description: xi, 227 p.: ill (some col.), col. maps; 27 cm. ISBN: 9781608762033 (hardcover) 1608762033 (hardcover) LC classification: HM1271 .C8522 2010 Related names: Kerwin, Lydia B. Contents: Language identity in a multicultural Canada: the case of deaf Canadians / Campbell McDermid -- The protection and promotion of cultural diversity at the international level / Mira Burri-Nenova -- Cultural diversity, immigrants and international trade: some empirical observations from nine OECD countries / Roger White and Bedassa Tadesse -- Narrative as methodology: understanding cultural diversity through unfolding cultural tales / Jessie Bee Kim Koh -- Managing cultural diversity: competing discourses in Australian mulitculturalism / Justine Dandy -- The development of cultural diversity tourism: policy framework and options / Peter U.C. Dieke -- Families, education and cultural diversity: program designing and evaluation / M.A. Santos Rego, M. Lorenzo Moledo and A. Godás Othero -- Cross-cultural knowledge creation: a native American case study / Gelaye Debebe -- Parenthood and migration: cultural and psychological issues / Olivier Taïeb ... [et al.] -- Biocultural diversity conservation: management and recovery of rural Mediterranean landscapes / A.J. Rescia ... [et al.] -- Global economy and cultural diversity / Janez Prašnikar and Andreja Cirman. Subjects: Cultural pluralism. Notes: Includes bibliographical references and index. Series: Focus on civilizations and cultures series Focus on civilizations and cultures.

Culture, growth and economic policy / Panagiotis E. Petrakis. LCCN: 2014932054 Personal name: Petrakēs, P., author. Published/Produced: Heidelberg Springer, [2014] Description: x, 236 pages: illustrations; 24 cm Links: Table of contents only http://www.loc.gov/catdir/enhancements/fy1412/20149320 54-t.html Publisher description http://www.loc.gov/catdir/enhancements/fy 1412/2014932054-d.html Contributor biographical information http://www. loc.gov/catdir/enhancements/fy1412/2014932054-b.html ISBN: 3642414397 9783642414398 LC classification: HM548 .P47 2014 Contents: Machine generated contents note: 1.Introduction -- References -- pt. I The Foundations -- 2.Human Needs and Cultural Background -- 2.1.Human Needs -- 2.2.Needs and Economic Evolution -- 2.3.Needs, Economic Growth and Cultural Background -- 2.4.From Individual to Aggregate Level -- References -- 3.The

11.3.Culture, Values and Their Diachronic Development -- 11.4.Differences in Values from Generation to Generation -- 11.5.The Preservation of Religious Beliefs Over Time -- References -- pt. III Cultural Background, Growth and Policy -- 12.Economic Growth and Cultural Background -- 12.1.Data -- 12.2.Methodology Employed and Empirical Results -- 12.3.Empirical Work and Discussion -- 12.3.1.The Basic Model -- 12.3.2.Sensitivity Analysis -- 12.3.3.The Effects of Culture on Growth and Cultural Change -- References -- 13.Cultural Background, Transaction Costs and Economic Institutions -- 13.1.The Role of Transaction Costs -- 13.2.The Variables Used -- 13.3.The Methodology Employed -- 13.4.Empirical Work and Discussion -- 13.5.The Effects of Culture in Shaping Transactions and Economic Institutions -- References -- 14.The Effects of Culture, Transactions and Institutions on Opportunity Entrepreneurship -- 14.1.The Theoretical Background and the Variables Used -- 14.2.The Endogenous Problem -- 14.3.Hypotheses Testing -- 14.4.The Data -- 14.5.The Methodology Employed -- 14.6.Empirical Results and Discussion -- 14.7.Overall Relationships Derived and Policy Implications -- References -- 15.The Effects of Cultural Background and Knowledge Creation on Self-Employment and Entry Density Rates -- 15.1.Entrepreneurship Identification, Knowledge and Cultural Background -- 15.2.The Data -- 15.3.The Methodology Employed -- 15.4.Empirical Work and Discussion of the Results -- 15.5.The Overall Relationships Derived -- References -- 16.Cultural Background, Growth and Economic Policy -- 16.1.The Middle-Income Trap -- 16.2.Why Do Countries Fall into the Middle-Income Trap? An Economic Explanation -- 16.3.The Cultural Growth Trap -- 16.4.Ideal Cultural Background and Growth -- 16.5.Austerity Policy and Cultural Requirements -- 16.6.A Neo-Keynesian Framework of Production and Employment Decisions and Cultural Requirements -- 16.7.Stabilization Policy and Cultural Background in South Europe -- References -- Appendix -- Appendix A -- Appendix B -- Appendix C -- Appendix D. Subjects: Culture--Economic aspects. Economic development. Economic policy. Notes: Includes bibliographical references.

Expect more: demanding better libraries for today's complex world / R. David Lankes. LCCN: 2013375159 Personal name: Lankes, R. David, 1970- Published/Created: [S.l.]: R. David Lankes, c2012. Description: vi, 118 p.: ill.; 23 cm. ISBN: 9781477476352 (pbk.) 1477476350 (pbk.) LC classification: Z665 .L37 2012 Summary: "Libraries have existed for millennia, but today many question their necessity. In an ever more digital and connected world, do we still need places of books in our towns, colleges, or schools? If libraries aren't about books, what are they about? In Expect More, David Lankes, winner of the 2012 ABC-CLIO/Greenwood Award for the Best Book in

Library Literature, walks you through what to expect out of your library. Lankes argues that, to thrive, communities need libraries that go beyond bricks and mortar, and beyond books and literature. We need to expect more out of our libraries. They should be places of learning and advocates for our communities in terms of privacy, intellectual property, and economic development. Expect More is a rallying call to communities to raise the bar, and their expectations, for great libraries."--Author's website. Contents: 1. The Arab Spring: expect the exceptional -- 2. The argument for better libraries: expect impact: Collective buying agent; Economic stimulus; Center of learning; Safety net; Steward of cultural heritage; Cradle of democracy; Symbol of community aspirations -- 3. The mission of libraries: expect more than books: I love reading...no really; Mission to nowhere?; A mission based on higher expectations -- 4. Facilitating knowledge creation: expect to create: Library as facilitator; What is knowledge?; Expanding the definition of facilitation; Teacher, librarian, tinker, spy -- 5. Improve society: expect grander: Expecting more than pie and prostitutes; Of the community; Walled gardens; Grand challenges; Is my library that grand? -- 6. Communities: expect a platform: Library as platform; Libraries as place -- 7. Librarians: expect brilliance: Librarian by hire; Librarian by degree; Librarian by spirit; Salzburg and a few of my favorite things; The facilitators; Adding up a librarian -- 6. Action plan: expect more: Action plan for great libraries; Action plan for bad libraries; Action plan for good libraries. Subjects: Library science--Philosophy. Library science--Forecasting. Libraries and community. Libraries and society. Notes: Also available in electronic format. Includes bibliographical references and index.

Global leadership: research, practice, and development / Mark E. Mendenhall ... [et al.]. LCCN: 2012010533 Edition: 2nd ed. Published/Created: New York: Routledge, 2013. Description: xii, 308 p.: ill.; 27 cm. ISBN: 9780415808859 (hardback: alk. paper) 9780415808866 (pbk.: alk. paper) 9780203138014 (ebook) LC classification: HD57.7 .G6527 2013 Related names: Mendenhall, Mark E., 1956- Contents: Leadership and the birth of global leadership / Mark E. Mendenhall -- The multidisciplinary roots of global leadership / Joyce S. Osland -- An overview of the global leadership literature / Joyce S. Osland -- Mapping the content domain of global leadership competencies / Allan Bird -- Process models of global leadership development / Joyce S. Osland, Allan Bird -- Assessing global leadership competencies / Allan Bird, Michael J. Stevens -- Leading global teams / Martha Maznevski, Celia Chui -- Global leadership knowledge creation and transfer / Allan Bird, Gary R. Oddou -- Leading global change / Joyce S. Osland -- Global leadership development / Gary R. Oddou, Mark E. Mendenhall -- Responsible global leadership /

Günter K. Stahl, Nicola M. Pless, Thomas Maak -- Looking to the future / B. Sebastian Reiche, Mark E. Mendenhall. Subjects: Leadership. Executives--Training of. International business enterprises--Personnel management. Notes: Includes bibliographical references (p. [269]-302) and index. Series: Routledge global human resource management series

Guided inquiry design in action: middle school / Leslie K. Maniotes, LaDawna Harrington, and Patrice Lambusta; foreword by Carol C. Kuhlthau. LCCN: 2015022398 Personal name: Maniotes, Leslie K., 1967- Published/Produced: Santa Barbara, CA: Libraries Unlimited, 2016. Description: xiv, 158 pages: illustrations; 28 cm. ISBN: 9781440837647 (paperback) 9781440837654 (ebook) LC classification: LB1060 .M327 2016 Related names: Harrington, LaDawna. Lambusta, Patrice. Summary: "Supplying classroom-tested lessons and unit plans that can serve as templates, this book demonstrates exactly how to integrate and implement Guided Inquiry Design (GID) theory into practice. Answers the needs of teachers and librarians who are seeking actual lesson plans using the GID concepts specifically at the 6th-8th grade levels. Supplies lesson plans and complete units of Guided Inquiry Design along with materials for implementation. Includes techniques for assessment of learning strategies aligned to the Common Core State Standards. Serves to heighten student engagement by going beyond fact-finding to achieve deeper understanding and knowledge creation. Encourages embedded information literacy and creates student choice"-- Provided by publisher. "Are you looking for examples of Guided Inquiry in real-world practice that can be applied to your specific learning context? This book shows you how fully designed units have been successfully executed in the middle-school context"-- Provided by publisher. Contents: Machine generated contents note: -- CONTENTS Foreword by Carol C. Kuhlthau Acknowledgments Introduction Chapter 1: Guided Inquiry in the Middle School Chapter 2: Guided Inquiry Design Framework Chapter 3: Schedules, Routines, and Strategies for Inquiry in Middle School Chapter 4: Inquiry Tools for Deep Learning Chapter 5: Listen My Children: A Unit Blending History and Poetry Chapter 6: We're Coming to America: A Unit that Inspires Empathy and Understanding Chapter 7: True Detectives: An Interdisciplinary Unit Chapter 8: Celebrate Life!: A Unit About What Makes People Great Chapter 9: Getting Started and Sustaining Change References Index . Subjects: Learning. Motivation in education. Information literacy--Study and teaching (Middle school) Middle school teaching. Education / Professional Development. Language Arts and Disciplines / Literacy. Notes: Includes bibliographical references (pages 153-154) and index. Series: Libraries Unlimited guided inquiry

Handbook of manufacturing industries in the world economy / edited by John R. Bryson, Professor of Enterprise and Competitiveness, Birmingham BUsiness School, University of Birmingham, UK, Jennifer Clark, Associate Professor, School of Public Policy, Georgia Institute of Technology, USA, Vida Vanchan, Associate Professor, Geography and Planning Department, State University of New York (SUNY) Buffalo State, USA. LCCN: 2014950966 Published/Produced: Cheltenham, UK; Northampton, MA: Edward Elgar Publishing, [2015] ©2015. Description: xvii, 519 pages: black and white illustrations, maps and charts; 25 cm. ISBN: 1781003920 9781781003923 LC classification: HD9720.5 .H36 2015 Related names: Bryson, J. R., 1963- edt Clark, Jennifer. edt Vanchan, Vida. edt Contents: PART I: INTRODUCTION. Manufacturing Matters: Space, Place, Time and Production / Jennifer Clark, John R. Bryson and Vida Vanchan -- PART II:(PROCESSES) BUILDING BLOCKS: FACTOR INPUTS AND PRODUCTION ORGANIZATION. Manufacturing Management in Theory and Practice / Paul L. Forrester -- Manufacturing and Labor / Sally Weller -- How Does Financialization Affect Manufacturing Investment? Preliminary Evidence from the US and UK / Susan Christopherson -- Manufacturing Logistics? / Peter V. Hall -- Reshoring and the 'Manufacturing Moment' / Margaret Cowell and John Provo -- Relocation of Production Activities and Underlying Social Dynamics: An Analytical Framework based on a Canadian Perspective / Patrice Jalette -- Tool-less Manufacture: Digital Fabrication, 3D Printing and the Third Industrial Revolution / Michael Ward -- Engineering and Manufacturing: Concurrent Maturation of xRL / Ben Wang, William C. Kessler and Andrew Dugenske -- Energy and Manufacturing: Technology and Policy Transformations and Challenges / Marilyn A. Brown and Gyungwon Kim -- Design and Manufacturing: The Competitiveness of American, European and Chinese Industrial Design Companies / Vida Vanchan and John R. Bryson -- Intellectual Property and Patents: Knowledge Creation and Diffusion / Dieter F. Kogler -- PART III: INDUSTRY AND FIRM CASE STUDIES. Manufacturing Textile Futures: Innovation, Adaptation and the UK Textiles Industry / Megan Ronayne -- Finding a Future for the U.S. Furniture Industry / Susan Walcott -- New Geographies of Advanced Manufacturing: The Case of Machine Tools / Ronald V. Kalafsky -- Farm Machinery: A Changing Path to Feed the World / Dawn M. Drake -- Hidden in Plain Sight: The North American Optics and Photonics Industry / Jennifer Clark -- Traditional and Emerging Markets in the Global Steel Supply Chain / Carey Durkin Treado -- Intermediate Manufacturing: Profit, Dependency and Value Attainment in Supply Chains / Rachel Mulhall -- Aerospace Manufacturing: Past, Present and Future / Colin G. Drury -- Manufacturing Stoke: Emergence, Transformation and Consolidation in the Surfboard

Industry / Andrew Warren and Chris Gibson -- Migrant Manufacturing: Translocal Production and the Establishment of a Polish Bakery in Birmingham, UK / Catherine Harris -- Škoda Auto: The Transformation from a Domestic to a Tier Two Lead Firm / Petr Pavlínek -- Samsung: Restructuring, Innovation, and Global Networks / Sam Ock Park -- PART IV: POLICY NARRATIVES IN MANUFACTURING. Stability Amid Industrial Change: The Geography of U.S. Deindustrialization since 1980 / Marc Doussard and Greg Schrock -- Searching for Advanced Manufacturing in the United Kingdom and United States: Definitions, Measurement and Public Policy / Finbarr Livesey -- National Manufacturing Policy, Local Real Estate Markets, and the Missing Region: Prospects for Urban Industrial Development in the U.S. / Laura Wolf-Powers -- The City and Industry: Deurbanizing Manufacturing in New York City? / Lynn McCormick -- Manufacturing in the Knowledge Economy: Innovation in Low-tech Industries / Teis Hansen and Lars Winther -- Crafting a Comeback: Cultivating an Innovative Ecosystem in Mature Regions / Maryann Feldman and Lauren Lanahan -- From Skill Mismatch to Reinterpretation: Challenges and Solutions for Manufacturing Worker Retention and Recruitment / Nichola J. Lowe -- PART V: CONCLUSION. Regeneration Economies: Manufacturing as the Next Industrial Revolution / Jennifer Clark, John R. Bryson and Vida Vanchan. Subjects: Manufacturing industries. Industries. Industrieökonomie. Produktion. Form/Genre: Readings. Notes: Includes bibliographical references and index.

Innovation and health: theory, methodology and applications / Thomas Grebel. LCCN: 2011922871 Personal name: Grebel, Thomas, 1970- Published/Produced: Cheltenham, UK: Edward Elgar, [2011] Description: xi, 187 pages: illustrations; 24 cm. ISBN: 9780857932204 (hardback) LC classification: RA410.5 .G73 2011 Contents: Change in economics -- Health economics -- Towards a theory of innovation in health economics -- Knowledge creation in medicine -- Network evolution in medicine -- Technology diffusion in medicine. Subjects: Medical economics. Medical innovations. Notes: Includes bibliographical references (pages 163-179) and index. Series: Industrial dynamics, entrepreneurship and innovation

Innovation leadership: creating the landscape of health care / [edited by] Tim Porter-O'Grady, Kathy Malloch. LCCN: 2009024781 Published/Created: Sudbury, MA: Jones and Bartlett Publishers, c2010. Description: xv, 366 p.: ill.; 23 cm. ISBN: 9780763765439 (pbk.) LC classification: RA971 .I475 2010 Related names: Porter-O'Grady, Timothy. Malloch, Kathy. Contents: Leadership for innovation: from knowledge creation to transforming health care / Tim Porter-O'Grady -- Creating the organizational context for innovation / Kathy Malloch -- Innovation in action: a practical system for getting results / Scott Endsley -- Directed creativity: how to generate new ideas for transforming health care / Paul Plsek -- Transdisciplinary design and innovation in the classroom / Prasad Boradker -- Launch pad: creating the business case for innovation / Dan O'Neill -- Transforming healthcare quality through innovations in evidence-based practice / Bernadette Melnyk -- Combining diffusion of innovation, complex adaptive health care Organizations, and whole systems shared governance: 21st century alchemy / Greg Crow -- Leadership, innovation, and healing spaces -- Intuitive leader / Dominique Surel. Subjects: Health services administration. Leadership. Health Services Administration. Organizational Innovation. Leadership. Notes: Includes bibliographical references and index.

Innovation, technology and knowledge / edited by Charlie Karlsson, Börje Johansson and Roger R. Stough. LCCN: 2011016343 Published/Created: Abingdon, Oxon; New York: Routledge, 2012. Description: xix, 361 p.: ill., maps; 24 cm. ISBN: 9780415667784 (hb) 9780203615331 (eb) LC classification: HC240.9.E5 I66 2012 Related names: Karlsson, Charlie, 1945- Johansson, B. (Börje) Stough, Roger. Contents: Introduction / Charlie Karlsson, Börje Johansson and Roger R. Stough -- Three stories about national systems of innovation / Alina Lidén -- Innovation, productivity and export: evidence from Italy / Roberto Antonietti and Giulio Cainelli -- Profiting from the breakthrough: technology commercialisation in the global age / Samantha Sharpe -- Innovation and regional disparities-a survey of regional growth drivers and economic performance / Andreas P. Cornett and Nils Karl Sørensen -- Knowledge integration and network structure: modelling innovation activity / Piergiuseppe Morone and Richard Taylor -- Proximity, innovation and success within free/libre open source software development project / Nicola Bellantuono, Pierpaolo Pontrandolfo and Barbara Scozzi -- Knowledge creation and innovation in medium technology clusters / Riccardo Cappelin -- Invention, innovation and regional growth in Swedish regions / Olof Ejermo and Urban Gråsjö -- Intengible assets and MNEs' locational strategies for innovation-or, why the regional matters: empirical insights from Germany and the UK / Jan-Philipp Kramer ... [et al.] -- High-technology and regions in an era of open innovation / Darrene Hackler -- Regional innovation, growth and convergence-divergence patterns in the EU / George Petrakos and George Anastasiou -- A social capital approach to regional transformation / Marina Jogmark -- A multi-dimensional perspective on the role of social capital in innovation / Soogwan Doh -- The potential of place: using location to attract global talent / Kathrine Richardson, Richard

Florida and Kevin Stolarick. Subjects: Economic development--European Union countries. Technological innovations--European Union countries. Notes: Includes bibliographical references and index. Series: Routledge studies in global competition; 55

Integral green Zimbabwe: an African phoenix rising / edited by Elizabeth Mamukwa, Ronnie Lessem, and Alexander Schieffer. LCCN: 2014016391 Published/Produced: Burlington: Gower, [2014] Description: xxxviii, 267 pages; 25 cm. ISBN: 9781472438195 (hardback: alk. paper) LC classification: HC910.Z9 I577 2014 Related names: Lessem, Ronnie, editor of compilation. Schieffer, Alexander, editor of compilation. Mamukwa, Liz, editor of compilation. Contents: About editors -- Contributor biographies -- Prologue -- Renewing a nation: the Zimbabwean phoenix is rising -- The phoenix rising: towards an integral green Zimbabwe / by Elizabeth Sarudzai Mamukwa -- Renewing Zimbabwe: from myths of de-colonisation to integral state / by Sabelo Ndlovu-Gatsheni -- Activating nature and community: establishing communtiy-based economic self-sufficiency -- The "Chinyika-model": a case of an integral community and rural self-sufficiency / by Paul Chidara Muchineripi and Steve Hwesa Masango Kada -- Community-technology-integration: ict based rural transformation in Chinyika and beyond / by Samuel Muchineripi Kundishora -- Kushanya Mumamisha: integral community activation via community based tourism / by Kennedy Mukuruwambwa Mandevani -- Catalysing renewal via culture and spirituality: towards a culture-based developmental economy -- Amakhosi: theatre for community action / by Cont Mhlanga -- Nzuri: drawing on the rhythm of Africa / by Kariamu Welsh -- Goko Routungamiri: the gene of leadership / by Ezekiah Chasamhuka Benjamin -- Afrintuneurship: towards integral African enterprise development / by Tapuwa Sherekete Rushesha and Winnie Ndakaiteyi Mhaka -- Enabling innovation driven research: conceiving of a socially based knowledge economy -- Utungamiriri Uzere: Áintegral Ubuntu leadership as caring for society / by Passmore Musungwa Matupire -- African knowledge rhythms: the calabash of organisational knowledge creation / by Elizabeth Sarudzai Mamukwa -- Zimbabwean industrial ecology: weaving the web of technological and social innovation / by Joshua Mazorodze Chinyuku -- Transforming education, learning and enterprise: sustainable development via a living life-based economy -- African holistic management: renewing the soil as economic foundation / by Allan Savory -- Permaculture: nature as a guide for community learning and sustainable livelihoods / by Mark Kenneth "Kudakwashe" Marombedza -- Co-creating Chinyika valley: towards a rural "integral green" economy powered by mobile communication / by Jimmy "Mukundi" Shindi -- Co-evolving integral green

Zimbabwe: the Zimbabwean eagle, flying in the sky -- "CARE 4 Zimbabwe": towards a pundutso centre for integral development / by Alexander "Mukanya" Schieffer and Ronnie "Samanyanga" Lessem. Subjects: Sustainable development--Zimbabwe--Planning. Economic development--Technological innovations--Zimbabwe. Economic development--Social aspects--Zimbabwe. Zimbabwe--Social policy. Notes: Includes bibliographical references and index. Series: Integral green society and economy

Knowledge and systems science: enabling systemic knowledge synthesis / Yoshiteru Nakamori. LCCN: 2013008341 Personal name: Nakamori, Yoshiteru. Published/Produced: Boca Raton: CRC Press, [2014] Description: xiii, 234 pages: illustrations; 25 cm ISBN: 9781466593008 (hardback) LC classification: HM651 .N35 2014 Summary: "Knowledge science is a problem-oriented interdisciplinary field that explores the modeling and application of the knowledge creation process. It encompasses disciplines such as knowledge management, technology and service management, knowledge synthesis and creation, and innovation theory, with the aim of constructing a better knowledge-based society. Written by one of the leading experts in the field, this book introduces the theoretical and methodological foundations of knowledge science. It explores the human dimension in problem solving as well as the various cultural dimensions of knowledge science"-- Provided by publisher. Subjects: Knowledge, Sociology of. Thought and thinking. Knowledge management. Creative ability. Computers / Database Management / Data Mining Computers / Software Development and Engineering / Systems Analysis and Design TECHNOLOGY and ENGINEERING / Electronics / General Notes: "A Chapman and Hall Book." Includes bibliographical references (pages 221-229) and index. Includes bibliographical references and index.

Knowledge creation in education / Seng Chee Tan. LCCN: 2014942070 Personal name: Tan, Seng Chee. Published/Produced: New York: Springer, 2014. Links: Contributor biographical information http://www.loc.gov/catdir/enhancements/fy1411/2014942070-b.html Publisher description http://www.loc.gov/catdir/enhancements/fy1411/2014942070-d.html Table of contents only http://www.loc.gov/catdir/enhancements/fy1411/2014942070-t.html ISBN: 9789812870469

Knowledge creation process in science: key comparative findings from the Hitotsubashi-NISTEP-Georgia Tech scientists' survey in Japan and the US = Kagaku ni okeru chishiki seisan purosesu: Nichi-Bei no kagakusha ni taisuru daikibo chōsa kara no shuyō na hakken jijitsu / Nagaoka Sadao, Igami Masatsura, John P. Walsh, Ijichi Tomohiro. LCCN: 2013395263 Personal name: Nagaoka, Sadao, 1952- Published/Produced: Tōkyō-to Chiyoda-ku: Monbu Kagakushō Kagaku Gijutsu Seisaku Kenkyūjo, 2011. Description: ii, 106 pages: illustrations; 30 cm. LC classification: Q180.J3 N26 2011 Parallel title: Kagaku ni okeru chishiki seisan purosesu Subjects: Research--Japan. Research--United States. Notes: Cover title. Includes bibliographical references (page 106). Summary in Japanese. Series: Chōsa shiryō; 203

Knowledge Governance: Reasserting the Public Interest / Edited by Leonardo Burlamaqui, Ana Celia Castro and Rainer Kattel. LCCN: 2012027468 Published/Produced: London SE1 8HA, UK: ANTHEM PRESS, 2012. Description: xxix, 269 pages: illustrations; 25 cm ISBN: 9780857285355 (hbk: alk. paper) 0857285351 (hbk.: alk. paper) LC classification: HD30.2 .K63636 2012 Related names: Burlamaqui, Leonardo, editor of compilation. Castro, Ana Célia, editor of compilation. Kattel, Rainer, 1974- editor of compilation. Contents: List of abbreviations -- List of tables and figures -- Foreword / Richard Nelson -- Introduction / Leonardo Burlamaqui, Ana Celia Castro and Rainer Kattel -- Knowledge governance: building a framework -- Knowledge governance: an analytical approach and its policy implications / Leonardo Burlamaqui -- From intellectual property to knowledge governance: a micro-founded evolutionary explanation / Annalisa Primi -- Catching up and knowledge governance / Rainer Kattel -- Innovation, competition policies and intellectual property: institutional fragmentation and the case for better coordination -- Where do innovations come from?: transformations in the US economy, 1970-2006 / Fred Block and Matthew R. Keller -- Antitrust and intellectual property: conflicts and convergences / Mario Luiz Possas and Maria Tereza Leopardi Mello -- The politics of pharmaceutical patent examination in Brazil / Ken Shadlen -- Going forward: towards a knowledge governance research agenda -- Varieties of latin-american patent offices: comparative study of practices and procedures / Ana Celia Castro, Ana María Pacón and Monica Desiderio -- An interoperability principle for knowledge creation and governance: the role of emerging institutions / John Wilbanks and Carolina Rossini -- The search for alternatives to patents in the twenty-first century / Luigi Palombi. Subjects: Knowledge management. Notes: This title is also available as an eBook. Includes bibliographical references.

Knowledge growth and the allocation of time [electronic resource] / Robert E. Lucas, Jr., Benjamin Moll. LCCN: 2011657380 Personal name: Lucas, Robert E. Published/Created: Cambridge, MA: National Bureau of Economic Research, c2011. Links: http://www.nber.org/papers/w17495 LC classification: HB1 Related names: Moll, Benjamin. National Bureau of Economic Research. Abstract: "We analyze a model economy with many agents, each with a different productivity level. Agents divide their time between two activities: producing goods with the production-related knowledge they already have, and interacting with others in search of new, productivity-increasing ideas. These choices jointly determine the economy's current production level and its rate of learning and real growth. Individuals' time allocation decisions depend on the knowledge distribution because the productivity levels of others determine their own chances of improving their productivities through search. The time allocations of everyone in the economy in turn determine the evolution of its knowledge distribution. We construct the balanced growth path for this economy, thereby obtaining a theory of endogenous growth that captures in a tractable way the social nature of knowledge creation. We also study the allocation chosen by an idealized planner who takes into account and internalizes the external benefits of search, and tax structures that implement an optimal solution. Finally, we provide two examples of alternative learning technologies, as concrete illustrations of other directions that might be pursued"--National Bureau of Economic Research web site. Notes: Title from PDF file as viewed on 1/3/2012. Includes bibliographical references. Computer file info: System requirements: Adobe Acrobat Reader. Mode of access: World Wide Web. Additional formats: Also available in print. Series: NBER working paper series; working paper 17495 Working paper series (National Bureau of Economic Research: Online); working paper no. 17495.

Knowledge management and competitive advantage: issues and potential solutions / Michael A. Chilton and James M. Bloodgood. LCCN: 2013025627 Published/Produced: Hershey: Information Science Reference, [2014] ©2014 Description: xxii, 363 pages; 29 cm ISBN: 9781466646797 (hardcover) LC classification: HD30.2 .K636842196 2014 Related names: Chilton, Michael A., 1950- editor of compilation. Bloodgood, James M., 1962- Summary: "This book examines current research in support of knowledge management by focusing on how knowledge resources can be used to create and sustain competitive advantages, combining imitation and innovation theories"--Provided by publisher. Contents: How knowledge creation capabilities lead to competitive advantage -- Exploration of soft computing approaches in itemset mining -- A synthesis of organizational learning and knowledge management literatures -- Group's knowledge management and the role of eWOM: recommendations for business organizations -- Knowledge and its effects on external relationships with customers -- The role of social intelligence in acquiring external knowledge for human capital development, organisational learning and innovation -- Innovation in the process of sales and marketing: new experiences based on tablets -- Innovation in the process of sales and marketing: new experiences based on tablets -- Reactive and proactive dynamic capabilities: using the knowledge chain theory of competitiveness -- Exploring knowledge types and use of knowledge protection in organizations -- Towards customer knowledge management (CKM): where knowledge and customer meet -- Police knowledge management strategy -- Knowledge management and the roles it plays in achieving superior performance -- A method for assessing knowledge loss risk with departing personnel -- Organizational barriers and facilitators in embedding knowledge strategy -- Knowledge development and protection as competitive advantage -- The human element of the knowledge worker: identifying, managing, and protecting the intellectual capital within knowledge management -- The effect of knowledge transfer motivation and receiver capability and motivation on organizational performance -- KM in higher learning institutions: using erg theory as a lens to understand the sharing of academic tacit knowledge: problems and issues in developing countries - perspectives from Vietnam -- Harnessing knowledge power for competitive advantage -- Competitive advantage and automated sharing of tacit knowledge. Subjects: Knowledge management. Competition. Technological innovations--Economic aspects. Notes: Includes bibliographical references (pages 304-350) and index.

Knowledge management in organizations: a critical introduction / Donald Hislop. LCCN: 2015473319 Personal name: Hislop, Donald. Edition: Third Edition Published/Created: Oxford: Oxford University Press, [2013]. Description: xx, 284 pages; 25 cm ISBN: 9780199691937 (pbk.) 0199691932 (pbk.) LC classification: HD30.2 .H575 2013 Contents: The contemporary importance of knowledge and knowledge management -- Epistemologies of knowledge in the knowledge management literature. The objectivist perspective on knowledge -- The practice-based perspective on knowledge -- An introduction to key concepts. What is knowledge management? -- Knowledge-intensive firms and knowledge workers -- Learning and knowledge management -- Knowledge creation and organizational unlearning. Nonaka and knowledge creation theory -- Unlearning and the deliberate loss of knowledge -- Socio-cultural issues related to managing and sharing knowledge. The influence of socio-cultural factors in motivating workers to participate in knowledge management initiatives -- Communities of practice -- Cross-community, boundary-spanning knowledge processes -- Power, politics, conflict, and knowledge process -- Information and communication technologies and knowledge management -- Facilitating knowledge management via the use of human resource management practices -- Leadership, organization culture management, and knowledge management -- Reflections on the topic of knowledge management. Subjects: Knowledge management. Knowledge management. Wissens management. Notes: Previous edition: 2009. Includes bibliographical references (pages 259-277) and index.

Knowledge management: an introduction / Kevin C. Desouza and Scott Paquette. LCCN: 2011017262 Personal name: Desouza, Kevin C., 1979- Published/Created: New York: Neal-Schuman Publishers, c2011. Description: xvi, 351 p.: ill.; 23 cm. ISBN: 9781555707200 (alk. paper) 1555707203 (alk. paper) LC classification: HD30.2 .D469 2011 Related names: Paquette, Scott. Contents: An introduction to knowledge management / Kevin C. Desouza -- The concept of knowledge / Peter Baloh with Kevin C. Desouza and Scott Paquette -- The concept of management / Jongmin T. Moon with Kevin C. Desouza -- Knowledge creation / Scott Paquette and Kevin C. Desouza -- Knowledge organization / Scott Paquette -- Knowledge transfer / Scott Paquette and Kevin C. Desouza -- Knowledge application / Chen Ye with Kevin C. Desouza and Scott Paquette -- Building global knowledge management systems / Kevin C. Desouza and Chris Rivinus -- Building the business case for knowledge management / Kevin C. Desouza -- Managing knowledge for organizational value / Scott Paquette and Kevin C. Desouza. Subjects: Knowledge management. Knowledge Management. Organizational Innovation. Notes: Includes bibliographical references and index.

Knowledge matters: the public mission of the research university / edited by Diana Rhoten and Craig Calhoun. LCCN: 2010022006 Published/Created: New York: Columbia University Press, c2011. Description: xvi, 539 p.; 25 cm. ISBN: 9780231151146 (cloth: alk. paper) 0231151144 (cloth: alk. paper) 9780231521833 (ebook) 0231521839 (ebook) LC classification: LB2326.3 .K63 2011 Related names: Rhoten, Diana. Calhoun, Craig J., 1952- Summary: Universities Are Changing Around The World. In China and Africa there is massive expansion, while many of America's greatest public universities are experiencing major budget cuts. In Latin America universities have been affected by dictatorships and privatization but are now growing in ways central to economic development. In Europe universities built as state institutions are being told to raise more money from private sources and are being reorganized so they will compete better in global rankings. In this context clarity about the public mission of universities is vital, yet it is lacking both outside and inside academia. When universities educate students, is this simply a private benefit because it advances their careers? Or is it a public good because informed citizens are integral to democracy and essential for national economic development? How important is equal opportunity? What are the effects of hierarchy? Who pays now and who will pay tomorrow? Should the results of academic research be private property for sale or openly available for public use? Who sets the university research agendas? What kinds of scholarship flourish and what kinds suffer? Should producing competitive research take priority over educating competent students? Do international rankings distort these and other university priorities or provide needed objective assessments? What are the university's roles and responsibilities in terms of knowledge creation and dissemination today? And tomorrow? In this collection, scholars report from Asia, Africa, Europe, Latin America, and North America. They confront the realities and challenges of higher education as it is torn between multiple public and private agendas. This comparative perspective illuminates both the continuing importance of the university's public mission and the pressing need to clarify it. Diana Rhoten is the founder and director of the Knowledge Institutions Program and the Digital Media and Learning Project at the Social Science Research Council. She has published in a range of academic journals and advises cultural, scientific, and educational institutions on issues of organizational design, creative collaboration, and adaptive change. Craig Calhoun is president of the Social Science Research Council and University Professor of the Social Sciences at New York University. He has served in a variety of academic leadership positions, including as a dean, and has conducted research in many international settings. His most recent book is an edited collection, Robert K Merton: Sociology of Science and Sociology as Science (Columbia). --Book Jacket. Contents: The public mission of the research university / Craig Calhoun -- Great expectations, past promises, and golden ages: rethinking the "crisis" of public research universities / Gustavo E. Fischman, Sarah E. Igo, Diana Rhoten -- "El central volumen de la fuerza": global hegemony in higher education and research / Simon Marginson, Imanol Ordorika -- The state, the university, and society in Soviet and Russian higher education: the search for a new public mission / Mark S. Johnson, Andrey V. Kortunov -- Public research universities in Latin America and their relation to economic development / Juan Carlos Moreno-Brid, Pablo Ruiz-Nápoles -- When neoliberalism colonizes higher education in Asia: bringing the "public" back to the contemporary university / Ka Ho Mok -- Challenges for higher education in Africa, Ubuntu, and democratic justice / Yusef Waghid -- The idea of the public university and the national project in Africa: toward a full circle, from the 1960s to the present / N'Dri T. Assié-Lumumba, Tukumbi Lumumba-Kasongo -- Rethinking what is made public in the university's public mission / John Willinsky -- Public research universities: from land grant to federal grant to patent grant institutions / Diana Rhoten, Walter Powell -- German universities in the new knowledge ecology: current changes in research conditions and university-industry relations / Stefan Lange, Georg Krücken -- The micropolitics of knowledge in England and Europe: the Cambridge University IPRs controversy and its macropolitical lessons / Voldemar Tomusk -- Playing the quality game: whose quality and whose higher education? / John Brennan, Mala Singh -- The academic workplace: what we already know, what we still do not know, and what we would like to know / Christine Musselin -- Cultural formations of the public university: globalization, diversity, and the state at the University of Michigan / Michael D. Kennedy. Subjects: Education, Higher--Research. Public universities and colleges--Research. Education, Higher--Aims and objectives. Education, Higher--Social aspects. Knowledge, Sociology of. Education and globalization. Notes: "A Columbia/SSRC Book." Includes bibliographical references and index.

Knowledge science: modeling the knowledge creation process / editor, Yoshiteru Nakamori. LCCN: 2011027657 Published/Created: Boca Raton, FL: Taylor and Francis, 2011. Description: xii, 177 p.: ill.; 25 cm. ISBN: 9781439838365 (hardback) LC classification: QA76.9.D343 K57 2011 Related names: Nakamori, Yoshiteru. Subjects: Data mining. Knowledge acquisition (Expert systems) Knowledge management--Data processing. Business and Economics / Operations Research Computers / Database Management / Data Mining Computers / Software Development and Engineering / Systems Analysis and Design Notes: Includes bibliographical references (p. 161-171) and index.

Macrowikinomics: rebooting business and the world / Don Tapscott and Anthony D. Williams. LCCN: 2010023338 Personal name: Tapscott, Don, 1947- Published/Created: New York: Portfolio/Penguin, 2010. Description: viii, 424 p.; 24 cm. ISBN: 9781591843566 LC classification: HC79.I55 T369 2010 Variant title: Macro Wikinomics Related names: Williams, Anthony D., 1974- Contents: I From wikinomics to macrowikinomics: -- Rebooting the world -- Five principles for the age of networked intelligence -- II Rethinking the fundamentals: -- Opening up the financial services industry -- Bootstrapping innovation and wealth creation -- III Reindustrializing the planet: -- Reversing the tide of disruptive climate change: a new global power emerges -- Wikinomics meets the green energy economy -- The transportation revolution: moving around in the twenty-first century -- IV Learning, discovery, and well-being: -- Rethinking the university: collaborative learning -- Science 2.0: igniting knowledge creation in a networked world -- Collaborative health care -- V Turning the media inside out: -- The demise of the newspaper and the rise of the new news -- Inside the future of music: prosumers take center stage -- The future of television and film: just another cool internet application? -- VI Rebooting the public square: -- Creating public value: government as a platform for social achievement -- The rise of the citizen regulator -- Solving global problems: beyond the nation-state -- Fighting for justice: user-generated freedom from Tehran to Rangoon to Beijing -- VII Conclusion: -- Ground rules for reinvention: making wikinomics happen in your organization -- Leadership for a changing world. Subjects: Information technology--Economic aspects. Information technology--Social aspects. Online social networks--Economic aspects. Business networks--Social aspects. Technological innovations--Social aspects. Notes: Includes bibliographical references (p. [381]-409) and index.

Organizational knowledge dynamics: managing knowledge creation, acquisition, sharing, and transformation / Constantin Bratianu, Bucharest University of Economic Studies, Romania. LCCN: 2015006749 Personal name: Brătianu, Constantin, 1946- Published/Produced: Hershey, PA: Information Science Reference, An Imprint of IGI Global, [2015] Description: xxi, 349 pages; 29 cm ISBN: 9781466683181 (hardcover) LC classification: HD58.82 .B74 2015 Summary: "This book introduces the idea that organizational knowledge is composed of three knowledge fields: cognitive knowledge, emotional knowledge, and spiritual knowledge and is intended for readers interested in knowledge management, intellectual capital, human resources management, change management, and strategic management"-- Provided by publisher. Subjects: Organizational learning. Knowledge management. Notes: Includes bibliographical references (pages 307-345) and index. Series: A volume in the advances in knowledge, acquisition, transfer, and management (AKATM), book series

Puting knowledge to work: from knowledge transfer to knowledge exchange / edited by Amare Desta, Mentesnot Mengesha and Mammo Muchie. LCCN: 2016009850 Published/Produced: Trenton, New Jersey: Africa World Press, 2016. ISBN: 9781569024805 (pb) 9781569024799 (hb: alk. paper) LC classification: DT14 .P88 2016 Other title: Title should read: Putting knowledge to work Related names: Desta, Amare, 1964- editor. Mengesha, Mentesnot, 1959- editor. Muchie, Mammo, editor. Global Knowledge Exchange Network. Conference (1st: 2011: London, England) Global Knowledge Exchange Network. Conference (2nd: 2012: London, England) Contents: Linking indigenous knowledge with scientific knowledge -- Indigenous knowledge in the context of natural resource management: an information systems perspective / Amare Desta -- The role of diaspora in knowledge exchange network for national growth and social change / Tewabech Bishaw -- Indigenous knowledge systems (IKS): application and their impact on industrial activities of companies in Africa / Medicine Magocha -- Indigenous knowing and everyday challenges -- Mixing indigenous with exogenous knowledge in the formulation of agricultural policy in Ethiopia: the principle of "a foot on the ground" / Demese Chanyalew -- Information for micro decisions and livelihood improvement: outreach as a key challenge in Ethiopia / Bedru B. Balana -- Increasing water availability, quality and quantity through public expenditure: an experience from the public service delivery in Uganda / Samuel Galiwango -- Putting research knowledge into action: the missing link for sustainability of Lake Tana ecosystem, Ethiopia / Dessalegn Molla Ketema, Nick Chisholm and Pat Enright -- Economic contribution of pastoral and agro pastoral production to food security and livelihoods systems in Africa: the case of Eastern Sudan, Eritrea and Western Ethiopia in the Horn of Africa / Zeremariam Fre and Bereket T. Tsegay -- Comprehending globalization and local development / Amare Desta and Mammo Muchie -- Indigenous knowledge in compatible with scientific knowledge -- The nano-revolution and economic development: is Africa ready? / Hailemichael Teshome and Mammo Muchie -- The implications of problems of good governance in FDI flow and transfer of knowledge to Ethiopia / Semahagn Gashu -- When lightweights beat heavyweights / Tadesse Biru Kersmo -- What could have predicted the North African and the Middle East uprising? / Yared Haile-Meskel -- Exploring governance of Lake Tana fishery: interactive perspective on governance / Dessalegn Molla Ketema, Nickolas Chisholm and Patrick Enright -- Revisiting the role of indigenous knowledge in higher learning institution -- Promises and perils of the virtuous knowledge exchange cycle in Ethiopian HEIs / Elizabeth Ayalew -- Prospects of knowledge sharing among Ethiopian institutions of higher learning / Rahel Bekele and Ermias Abebe -- Potentials of indigenous techniques of knowledge creation in Qinea schools of Ethiopia / Mariye Yigzaw Workineh -- Intellectual property rights: a focus on medicinal plant products from Africa / Fekadu Fullas and Mammo Muchie -- Conclusion. Subjects: African diaspora--Intellectual life--Congresses. Knowledge management--Africa--Congresses. Technology transfer--Africa--Congresses. Africa--Intellectual life--Congresses. Notes: Selected papers from the 1st and 2nd yearly conferences of GKEN (Global Knowledge Exchange Network), held 2011 and 2012 in London, England. Includes bibliographical references and index.

Science, philosophy and sustainability: the end of the Cartesian dream / edited by Ângela Guimarães Pereira, Silvio Funtowicz. LCCN: 2014034419 Published/Produced: London; New York: Routledge, Taylor and Francis Group, 2015. Description: xxi, 169 pages; 24 cm. Links: Cover image http://images.tandf.co.uk/common/jackets/websmall/978113879/9781138796409.jpg ISBN: 9781138796409 (hbk) 1138796409 (hbk) LC classification: Q175.5 .S3655 2015 Related names: Pereira, Ângela Guimarães. Funtowicz, Silvio. Summary: "This book proposes to consider new ways in which science can be used to sustain our planet and enrich our lives. For science to remain a legitimate and trustworthy source of knowledge, society will have to engage in the collective processes of knowledge co-production, which not only includes science, but also other types of knowledge. This process of change has to include a new commitment to knowledge creation and transmission and its role in a plural society. Social responsibility has historically played an important role in science and engineering, but has lost momentum nowadays because many facets of science are lost due to reduction caused by the hybridization of science and policy. This book helps to release and reactivate social responsibility within contemporary science and technology"-- Provided by publisher. Subjects: Science--Social aspects. Science--Philosophy. Technology--Philosophy. Business and Economics / Development / Sustainable Development. POLITICAL SCIENCE / Public Policy / Environmental Policy. Notes: "Earthscan from Routledge." Includes bibliographical references and index. Series: Routledge explorations in sustainability and governance

Social software and the evolution of user expertise: future trends in knowledge creation and dissemination / Tatjana Takseva, editor. LCCN: 2012019283 Published/Created: Hershey, PA: Information Science Reference, c2013. Description: xxx, 395 p.: ill.; 29 cm. ISBN: 9781466621787 (hardcover) 9781466621794 (ebook) 9781466621800 (print and perpetual access) LC classification: H62 .S72471817 2013 Related names: Takseva, Tatjana, 1970- Summary: "This book examines the vital role that social software applications play in regards to the cultural definitions by experts and challenges the reader to consider how recent changes in this area influence how we create and distribute knowledge"-- Provided by publisher. Subjects: Social sciences--Research--Methodology. Online social networks. Web 2.0 Notes: Includes bibliographical references and index.

Source: the inner path of knowledge creation / Joseph Jaworski; Betty Sue Flowers, editor. LCCN: 2011039994 Personal name: Jaworski, Joseph, 1934- Edition: 1st ed. Published/Created: San Francisco: Berret-Koehler Publishers, Inc., c2012. Description: x, 221 p.: ill.; 24 cm. ISBN: 9781576759042 (hbk.: alk. paper) LC classification: BJ1470 .J37 2012 Related names: Flowers, Betty S. Subjects: Self-realization. Creative ability. Leadership. Success.

Entrepreneurship--Psychological aspects. Notes: Includes bibliographical references (p. 201-210) and index.

Stories of innovation for the millennial generation: the Lynceus long view / Piero Formica. LCCN: 2013025083 Personal name: Formica, Piero. Edition: First edition. Published/Produced: New York, NY: Palgrave Macmillan, 2013. Description: xviii, 149 pages: illustrations; 23 cm. Links: Cover image http://www.netread.com/jcusers2/bk1388/084/9781137350084/image/lgcover. 9781137350084.jpg ISBN: 9781137350084 (Hardback) LC classification: HD45 .F664 2013 Summary: "The potential of storytelling as a research tool for enhancing the understanding of knowledge creation, acquisition and conversion into innovation and innovative business activities is the methodological underpinning of this book's narrative approach. The subtitle comes from Lynceus, one of the Argonauts who accompanied Jason to Colchis in his quest to find the Golden Fleece, who was said to have excellent sight. Among the various interpretations of the meaning of the Fleece, one version highlights the importance of discovery to innovation by voyaging to 'terrae incognitae' (unknown lands). This book is a narrative of a fictional voyage to the 'terra incognita' of Innoland - the island of innovation and other mental travels that make sense of events and actions which spur innovation.Stories of Innovation for the Millennial Generation is written for Millennials willing to assimilate and grow dynamic, innovation-driven capabilities which lead to the creation of high-impact startups"-- Provided by publisher. Contents: Machine generated contents note: -- Foreword -- Professor Martin Curley -- Preface -- Making Connections - Charles Dickens, 21st Century Innovation and the Magpie Tactic -- John Edmondson -- Acknowledgements -- Introducing the Stories -- Chapter 1: The Wonder Generation: Adolescent Entrepreneurs of Generation Y, New Stars of Innovation -- Chapter 2: The Lynceus Long View -- The Stories -- Chapter 3: Self-portrait with Thoughts of a Millennial, Youthful Knowledge Nomad -- Chapter 4: An Italian Millennial Fleeing His Country -- Chapter 5: Innovation: a Disquieting Muse -- Chapter 6: A Voyage to Innoland, Island of Innovation: Tales of a Modern Gulliver -- Chapter 7: In Praise of the Black Sheep: the Country That I Would Like -- Chapter 8: War and Peace: Destructive Creativity and Creative Destruction -- Chapter 9: Three Visitors in the Emerald City -- Chapter 10: The Expulsion of the Oblomovs: From the Cult of Concealment to the Practice of Innovation -- Chapter 11: In the Kingdom of Zenda: Prisoners of Success -- Stories from the 'Sea of Crises' -- Chapter 12: Mr X; or, the Anxieties of the Forgotten Man -- Chapter 13: Extracts of a Report of the Ambassador of Arithmos to Oeconomicus -- Chapter 14: 'On Beauty': Three Scherzos Played on the World Economy Stage

-- Chapter 15: The War of the Two 'F's and the Decisive Intervention of 'E' -- Chapter 16: Autobiographical Sketches of Kidnapped Europa and Titanic Asia: a Tale of the Future Foreseen -- Chapter 17: Innovation in the Academic World: a New Mission for University Old Stars -- Chapter 18: A dialogue between start-up entrepreneurs Crusoe and Gulliver: Gulliver teaches Crusoe the physics of entrepreneurship -- Chapter 19: The Spider's Webs of Mr F and Mr M -- Chapter 20: The Red Thread of Coffee -- Chapter 21: Subterranean Currents: Trade Relations Between the Middle Kingdom and Arabia Felix in the Digital Age -- Chapter 22: An Interview with Maison Dubai's Style Director -- Chapter 23: Letter to the Millennial Students of Tartu University: Back to the Future -- Epilogue: Learning How Events Not Supposed to Happen, Happen -- Bibliography (selected). Subjects: Technological innovations. Creative ability in business. Generation Y. Business and Economics / Entrepreneurship. Business and Economics / Management. Business and Economics / Workplace Culture. Notes: Includes bibliographical references (pages 146-149). Series: Palgrave pivot

Territorial patterns of innovation: an inquiry on the knowledge economy in European regions / edited by Roberta Capello and Camilla Lenzi. LCCN: 2012049035 Published/Produced: London: Routledge, 2013. Description: xi, 354 pages: illustrations; 24 cm. ISBN: 9780415636063 (hbk.) 041563606X (hbk.) LC classification: HD30.2 .T477 2013 Related names: Capello, Roberta, 1962- Lenzi, Camilla. Contents: An inquiry on the regional dimension of the knowledge economy: aims and novelties of the book / Roberta Capello and Camilla Lenzi -- Measuring the territorial dimension of knowledge and innovation in Europe -- Technologically advanced regions / Roberta Capello and Andrea Caragliu -- Scientific regions / Alessandra Colombelli, Marta Foddi and Raffaele Paci -- Knowledge networking regions / Ernest Miguélez, Rosina Moreno and Jordi Suriñach 5. The Knowledge-Innovation Nexus in European Regions / Roberta Capello and Camilla Lenzi -- Territorial patterns of innovation and regional dynamics in Europe -- Territorial patterns of innovation / Roberta Capello -- Territorial patterns of innovation: evidence from European regions / Roberta Capello and Camilla Lenzi -- Knowledge, Human Capital and Regional Performance / Marta Foddi, Emanuela Marrocu, Raffaele Paci and Stefano Usai -- Knowledge flows and regional knowledge creation / Ernest Miguélez, Rosina Moreno and Jordi Suriñach -- Knowledge, innovation and economic efficiency across territorial patterns of innovation / Roberta Capello and Camilla Lenzi -- Regional knowledge creation in China, India and the US: lessons for European regions / Riccardo Crescenzi and Andrés Rodríguez-Pose -- Innovation policy implications -- Towards a conclusion: smart innovation policies / Roberto Camagni and Roberta Capello. Subjects: Knowledge economy--Europe. Economic development--Europe. Technological innovations--Economic aspects--Europe. Notes: Includes bibliographical references (pages 327-349) and index. Series: Routledge advances in regional economics, science and policy

Testimony: a philosophical introduction / Joseph Shieber. LCCN: 2014046901 Personal name: Shieber, Joseph. Published/Produced: New York: Routledge, Taylor and Francis Group, 2015. Description: xi, 221 pages; 24 cm ISBN: 9780415821322 (hardback) 9780415821339 (paperback) LC classification: BD238.T47 S54 2015 Summary: "The epistemology of testimony has experienced a growth in interest over the last twenty-five years that can be matched by few, if any, other areas of philosophy. Testimony: A Philosophical Introduction is an introduction to the epistemology of testimony that surveys this rapidly growing research area while incorporating a discussion of relevant empirical work from social and developmental psychology, as well as from the study of knowledge creation in groups. The past decade has seen a number of books on the epistemology of testimony, but there is a dearth of books that survey the current field. This book fills that gap, assessing the strengths and weaknesses of all major competing theories. All chapters conclude with Suggestions for Further Reading and Discussion Questions"-- Provided by publisher. Subjects: Testimony (Theory of knowledge) PHILOSOPHY / General. Notes: Includes bibliographical references (pages 199-216) and index.

The complete guide to knowledge management: a strategic plan to leverage your company's intellectual capital / Edna Pasher and Tuvya Ronen. LCCN: 2010051230 Personal name: Pasher, Edna. Published/Created: Hoboken, N.J.: John Wiley, c2011. Description: xvii, 204 p.: ill.; 24 cm. Links: Cover image http://catalogimages.wiley.com/images/db/jimages/9780470881293.jpg ISBN: 9780470881293 (hardback) 0470881291 (hardback) LC classification: HD30.2 .P375 2011 Related names: Ronen, Tuvya. Summary: "A straightforward guide to leveraging your company's intellectual capital by creating a knowledge management culture The Complete Guide to Knowledge Management offers managers the tools they need to create an organizational culture that improves knowledge sharing, reuse, learning, collaboration, and innovation to ensure mesurable growth. Written by internationally recognized knowledge management pioneers, it addresses all those topics in knowledge management that a manager needs to ensure organizational success. Provides plenty of real-life examples and case studies Includes interviews with prominent managers who have successfully implemented knowledge management structures within their organizations Offers chapters composed of short theoretical explanations and practical methods that you can utilize, based primarily on hands-on author experience Taking an intellectual journey into knowledge management, beginning with an understanding of the concept of intellectual capital and how to establish an appropriate culture, this book looks at the human aspects of managing knowledge workers, promoting interactions for knowledge creation and sharing."-- Provided by publisher. Subjects: Knowledge management. Intellectual capital--Management. Notes: Includes bibliographical references and index.

The creative capital of cities: interactive knowledge creation and the urbanization economies of innovation / Stefan Krätke. LCCN: 2011013465 Personal name: Krätke, Stefan, 1952- Published/Created: Chichester, West Sussex; Malden, MA: Wiley-Blackwell, 2011. Description: ix, 249 p.: ill., maps; 24 cm. Links: Cover image http://catalogimages.wiley.com/images/db/jimages/9781444336214.jpg Table of contents only http://catdir.loc.gov/catdir/enhancements/fy1109/2011013465-t.html Contributor biographical information http://catdir.loc.gov/catdir/enhancements/fy1109/2011013465-b.html Publisher descripttion http://catdir.loc.gov/catdir/enhancements/fy1109/2011013465-d.html ISBN: 9781444336214 1444336215 9781444336221 (pbk.) 1444336223 (pbk.) LC classification: BF408 .K73 2011 Summary: "The Creative Capital of Cities challenges the new urban growth concepts of creative industries from a critical urban theory perspective. The author presents both technological creativity in knowledge-intensive industries and artistical creativity in the cultural economy sector in the analysis of the 'creative capital of cities'. Krätke develops a transdisciplinary approach to the analysis of creativity and knowledge generation in an urban context by combining perspectives of economic geography, regional research, and socio-cultural urban studies. Providing deep insights backed up by rigorous empirical analysis, The Creative Capital of Cities reveals the complex relational dimensions of creativity while offering an alternate path to the assessment of urban economic success"--Provided by publisher. Contents: 1. Creativity and innovation under the command of capital. -- The capitalist imperative of creativity and innovation. -- Generalizing models of urban economic development. -- The role of innovation and inter-urban competition in Harvey's theory of capitalist urbanization. -- 2. Creative cities as a new urban growth ideology. -- The impact of creative occupations on regional economic success. -- Critique of Florida's conception of the creative class. -- An assessment of regional economic success factors and the impact of creative workers on regional development. -- The impact of creative occupational groups on regional economic development in Germany. -- Relationship between "qualities of place" and the regional concentration of scientifically and technologically creative workers. 3. Innovation and knowledge networks in a metropolitan region. -- The impact of localization economies and networking on technological creativity. -- Innovation and knowledge networks: Theoretical approaches. -- The application of network analysis to urban regions" knowledge networks. -- Geographic scales and structural properties of knowledge networks in the metropolitan region of Hanover. -- Assessment of network impacts on regional firms" innovation output. -- 4. Creativity in the culture and media industries. -- The impact of commercial imperatives on artistic creativity. -- The institutional order of the cultural economy: Creativity in a capitalist context. -- Global centres of the culture industry and the production of lifestyle images. -- 5. Local clustering of the cultural economy in the metropolis of Berlin. -- The urbanization economies of artistically creative occupations. -- The rise of the cultural economy in Berlin's inner-city area. -- Creative cities and the role of the culture industries in urban economic and spatial development: Implications for urban regeneration. -- 6. Synthesis: The creative capital of cities. Subjects: Creative ability. Technological innovations. Urbanization. Cities and towns. Notes: Machine generated contents note: Introduction.1. Creativity and innovation under the command of capital. The capitalist imperative of creativity and innovation. Generalizing models of urban economic development. The role of innovation and inter-urban competition in Harvey's theory of capitalist urbanization.Conclusion.2. Creative cities as a new urban growth ideology. The impact of creative occupations on regional economic success. Introduction. Critique of Florida's conception of the creative class. An assessment of regional economic success factors and the impact of creative workers on regional development. The impact of creative occupational groups on regional economic development in Germany. Relationship between 'qualities of place' and the regional concentration of scientifically and technologically creative workers.Conclusion.3. Innovation and knowledge networks in a metropolitan region. The impact of localization economies and networking on technological creativity. Introduction. Innovation and knowledge networks: Theoretical approaches. The application of network analysis to urban regions' knowledge networks. Geographic scales and structural properties of knowledge networks in the metropolitan region of Hanover. Assessment of network impacts on regional firms' innovation output.Conclusion.4. Creativity in the culture and media industries. The impact of commercial imperatives on artistic creativity. Introduction. The institutional order of the cultural economy: Creativity in a capitalist context. Global centres of the culture industry and the production of lifestyle images.Conclusion.5. Local clustering of the cultural economy in the metropolis of Berlin. The urbanization economies of artistically creative occupations. Introduction. The rise of the cultural economy in Berlin's inner-city area. Creative cities and the role of the culture industries in urban economic and spatial development: Implications for urban regeneration.Conclusion.6. Synthesis: The creative capital of cities. References. Appendix. Index. Includes bibliographical references (p. [219]-233) and index. Series: Studies in urban and social change Studies in urban and social change.

The emergent knowledge society and the future of higher education: Asian perspectives / edited by Deane E. Neubauer. LCCN: 2011025872 Published/Created: New York, NY: Routledge, 2012. Description: xiii, 223 p.: ill.; 24 cm. ISBN: 9780415608695 (hardback) 9780203145906 (ebook) LC classification: LA1058 .E47 2012 Related names: Neubauer, Deane E. Summary: "The nature of higher education is by no means fixed: it has evolved over time; different models of higher education co-exist alongside each other at present; and, worldwide, there are demands for higher education to change to better help support economic growth and to better fit chagning social and economic circumstances. This book examines, from an Asian perspective, the debates about how higher education should change. It considers questions of funding, and of who will attend universities, and the fundamental question of what universities are for, especially as the three key funcations of universities - knowledge creation through research, knowledge dissemination through teaching and service, and knowledge conservation through libraries, the disciplinary structuring of knowledge and in other ways - are increasingly being carried out much more widely outside universities in the new "knowledge society". Throughout, the book discusses the extent to which the countries of East Asia are developing new models of higher education, thereby better preparing themselves for the "new "knowledge society", rather than simply following old Western models"-- Provided by publisher. "This book examines, from an Asian perspective, the debates about how higher education should change. It considers questions of funding, and of who will attend universities, and the fundamental question of what universities are for, especially as the three key functions of universities - knowledge creation through research, knowledge dissemination through teaching and service, and knowledge conservation through libraries, the disciplinary structuring of knowledge and in other ways - are increasingly being carried out much more widely outside universities in the new "knowledge society""-- Provided by publisher. Subjects: Education, Higher--Asia--Cross-cultural studies. Educational change--Asia--Cross-cultural studies. Education / Higher Education / Comparative Social Science / Ethnic Studies / General Notes: Includes bibliographical references and index. Series: Comparative development and policy in Asia; 10

The essentials of knowledge management / edited by John S. Edwards. LCCN: 2015029076 Published/Produced: Houndmills, Basingstoke, Hampshire; New York, NY: Palgrave Macmillan, 2015. Description: x, 372 pages; 23 cm. ISBN: 9781137552082 (hardback) LC classification: HD30.2 .E84 2015 Related names: Edwards, John S., 1952- editor. Summary: "This book reviews the field of Knowledge Management, taking a holistic approach that includes both "soft" and "hard" aspects. It provides a broad perspective on the field, rather than one based on a single viewpoints from Computer Science or Organizational Learning, offering a comprehensive and integrated conception of Knowledge Management. The chapters represent the best Knowledge Management articles published in the 21st century in Knowledge Management Research and Practice and the European Journal of Information Systems, with contributors including Ikujiro Nonaka, Frada Burstein, and David Schwarz. Most of the chapters contribute significantly to practise as well as theory. The OR Essentials series presents a unique cross-section of high quality research work fundamental to understanding contemporary issues and research across a range of Operational Research topics. It brings together some of the best research papers from the highly respected journals of the Operational Research Society, also published by Palgrave Macmillan. "-- Provided by publisher. Contents: Machine generated contents note: -- 1. Introduction: setting the scene -- PART I: FOUNDATIONS -- 2. Management knowledge and knowledge management: realism and forms of truth -- 3. The theoretical foundations of knowledge management -- PART II: STRATEGIC ISSUES -- 4. The knowledge-creating theory revisited: knowledge creation as a synthesizing process -- 5. Absorptive capacity: a proposed operationalization -- 6. Knowledge as a measurable object in business contexts: a stock-and-flow approach -- 7. A conceptual framework for unlearning in a homecare setting -- 8. A study of knowledge management enablers across countries -- PART III: UNDERSTANDING KNOWLEDGE TRANSFER/SHARING -- 9. Transfer of knowledge in knowledge management systems: unexplored issues and suggested studies -- 10. A measure of knowledge sharing behavior: scale development and validation -- PART IV: PEOPLE OR TECHNOLOGY ISSUES? -- 11. Reproducing knowledge: Xerox and the story of knowledge management -- 12. Managing large amounts of knowledge objects: cognitive and organisational problems -- PART V: NEWER TECHNOLOGICAL DEVELOPMENTS -- 13. Codifying collaborative knowledge: using Wikipedia as a basis for automated ontology learning -- 14. Deciding to use an enterprise wiki: the role of social institutions and scripts -- 15. The role of social networks in knowledge creation. Subjects: Knowledge management. Strategic planning. Business and Economics / Management. Business and Economics / Management Science. Business and Economics / Strategic Planning. Business and Economics / Information Management. Business and Economics / Knowledge Capital. Notes: Includes bibliographical references and index. Series: OR essentials series

The global governance of knowledge creation and diffusion / [edited by] Helge Hveem, Lelio Iapadre. LCCN: 2010051166 Published/Created: New York, NY: Routledge, 2011. Description: xix, 254 p.: ill.; 24 cm. ISBN: 9780415605748 (hardback) 9780203813683 () LC classification: LB2331.5 .G56 2011 Related names: Hveem, Helge. Iapadre, Lelio. Summary: "There is an increasing scientific and political sensibility for questions regarding the "governance of a knowledge society" and the societal benefits and problems of a "knowledge economy". The International Regulation of Knowledge utilizes case studies and a comparative country / thematic approach to prove a comprehensive survey of the regulation and governance of knowledge flow, research and innovation. By identifying activities creating new knowledge (e.g., education, migration), it demonstrates how knowledge regulation and diffusion works in practice and policy. The book will be of interest to students and scholars of globalization, international relations and policy regulation"-- Provided by publisher. "The Global Governance of Knowledge provides a survey and analysis of international agreements and institutions, global and regional, which regulate the creation and dissemination of knowledge"-- Provided by publisher. Subjects: Education, Higher--International cooperation. Education and globalization. Knowledge management. Technology transfer. Diffusion of innovations. POLITICAL SCIENCE / General Notes: Includes bibliographical references and index. Series: Routledge/garnet series

The handbook of global science, technology, and innovation / edited by Daniele Archibugi and Andrea Filippetti. LCCN: 2015004138 Published/Produced: Chichester, West Sussex, UK: John Wiley and Sons Inc., [2015] Links: Cover image http://catalogimages.wiley.com/images/db/jimages/9781118739068.jpg ISBN: 9781118739068 (hardback) LC classification: T175 .H35 2015 Related names: Archibugi, Daniele. Filippetti, Andrea, 1977- Summary: "Provides a state-of-the-art overview of science, technology, and innovation in the context of globalization and global policy"-- Provided by publisher. Contents: Machine generated contents note: Notes on Contributors viii Editors'

Filippetti 24 From Governmental Open Data Toward Governmental Open Innovation (GOI): A Global Perspective 508 Sabine Brunswicker and Jeremiah Johnson 25 Serendipity and Chance in Scientific Discovery: Policy Implications for Global Society 529 Donald Gillies 26 Global Climate Change and the Direction of Technological Change 544 Andrew Tylecote 27 Global Risks: Cause and Consequence of the New Interactions Between Science, Technology, and Society 562 Jean-Yves Heurtebise 28 Globalization, Regionalization, and Technological Change 579 Frederick Guy Index 000 Subjects: Research--Industrial--Handbooks, manuals, etc. Technology transfer--Handbooks, manuals, etc. Technology and state--Handbooks, manuals, etc. Science and state--Handbooks, manuals, etc. POLITICAL SCIENCE / International Relations / Trade and Tariffs. Notes: Includes bibliographical references and index. Additional formats: Online version: Handbook of global science, technology, and innovation Chichester, West Sussex, UK: John Wiley and Sons Inc., [2015] 9781118738962 (DLC) 2015014494 Series: Hgp - handbooks of global policy

The learning curve: how business schools are re-inventing education / Santiago Iñiguez de Onzoño LCCN: 2011043419 Personal name: Iñiguez de Onzoño, Santiago. Published/Created: Houndmills, Basingstoke, Hampshire; New York: Palgrave Macmillan, 2011. Description: xxi, 202 p.: ill.; 24 cm. Links: Cover image http://www.netread.com/jcusers2/bk1388/236/9780230280236/image/lgcover.9780230280236.jpg ISBN: 9780230280236 (hardback) 0230280234 (hardback) LC classification: HF1106 .I58 2011 Summary: "How do you create world-class educational institutions that are academically rigorous and vocationally relevant? Are business schools the blueprint for institutions of the future, or an educational experiment gone wrong? This is the first title in a new series from IE Business School, IE Business Publishing"-- Provided by publisher. Contents: Part I. The Trailblazers: 1. Why management matters; 2. Out of the crisis, confronting the critics; 3. Developing better managers -- Part II. Today's Pressing Challenges: 4. The changing landscape; 5. E-Learning; 6. International stakeholders; 7. Melting pots of knowledge -- Part III. The Next Learning Curve: 8. The purpose of business schools; 9. The students of tomorrow; 10. Faculty and knowledge creation: how does it work? Subjects: Business education. Business schools. Management--Study and teaching. Notes: Includes bibliographical references (p. 182-195) and index.

The new librarianship field guide / R. David Lankes. LCCN: 2015039943 Personal name: Lankes, R. David, 1970- author. Published/Produced: Cambridge, MA; London, England: The MIT Press, [2015] ISBN:

9780262529082 (pbk.: alk. paper) LC classification: Z665 .L375 2015 Contents: Librarianship full stop -- They named the building after us -- The mission of librarians -- Knowledge creation -- Facilitation -- Participatory systems -- Improve society -- Librarians -- Pragmatic utopias -- What is a library -- Saving money and the world -- A platform for knowledge development -- Fitting knowledge in a box -- Academic libraries -- School libraries -- Public libraries -- Engines of advancement -- Coda -- Excursus: from mission to missionary -- Facilitating new librarianship learning -- Observations from the field -- FAQs (frequently argued questions) -- Discussion points. Subjects: Library science--Philosophy. Libraries and community. Libraries and society. Librarians--Attitudes. Notes: Includes bibliographical references and index.

The Oxford handbook of human capital / edited by Alan Burton-Jones, J.-C. Spender; foreword by Gary Becker. LCCN: 2011411099 Published/Created: Oxford: Oxford University Press, c2011. Description: xxviii, 688 p.: ill.; 26 cm. ISBN: 9780199532162 (hbk.) 0199532168 (hbk.) LC classification: HD4904.7 .O96 2011 Portion of title: Human capital Related names: Burton-Jones, Alan. Spender, J. -C. Contents: Part I. The nature of human capital -- ch. 1. An economic perspective on the notion of 'human capital' / Margaret M. Blair -- ch. 2. A social perspective: exploring the links between human capital and social capital / Janine Nahapiet -- ch. 3. Global culture capital and cosmopolitan human capital: the effects of global mindset and organizational routines on cultural intelligence and international experience / Kok-Yee Ng, Mei Ling Tan, and Soon Ang -- ch. 4. Cognition and human capital: the dynamic interrelationship between knowledge and behavior / Rhett A. Brymer, Michael A. Hitt, and Mario Schijven -- ch. 5. A capital-based approach to the firm: reflections on the nature and scope of the concept of capital and its extension to intangibles / Peter Lewin -- Part II. Human capital and the firm -- ch. 6. Human capital and transaction cost economics / Nicolai J. Foss -- ch. 7. Human capital and agency theory / J.-C. Spender -- ch. 8. Human capital in the resource-based view / Jeroen Kraaijenbrink -- ch. 9. Human capital, entrepreneurship, and the theory of the firm / Brian J. Loasby -- ch. 10. The firm, human capital, and knowledge creation / Georg von Krogh and Martin W. Wallin -- Part III. Human capital and organizational effectiveness -- ch. 11. Human capital, HR strategy, and organizational effectiveness / Peter Boxall -- ch. 12. How organizations obtain the human capital they need / Monika Hamori, Rocio Bonet, and Peter Cappelli -- ch. 13. Aligning human capital with organizational needs / David P. Lepak, Riki Takeuchi, and Juani Swart -- ch. 14. Maximizing value from human capital / Russell Coff -- ch. 15. Accounting for human capital and organizational effectiveness / Robin Kramar, Vijaya Murthy, and James Guthrie. Part IV. Human capital interdependencies -- ch. 16. Interdependencies between people in organizations / Robert M. Grant and James C. Hayton -- ch. 17. Understanding interdependencies between human capital and structural capital: some directions from Kantian pragmatism / David O'Donnell -- ch. 18. The distributed and dynamic dimensions of human capital / Ikujiro Nonaka, Ryoko Toyama, and Vesa Peltokorpi -- ch. 19. Human capital and the organization-accommodation relationship / Jacqueline C. Vischer -- ch. 20. Interdependencies between people and information systems in organizations / Alan Burton-Jones and Andrew Burton-Jones -- Part V. Human capital in the future economy -- ch. 21. Human capital, capabilites, and the firm: literati, numerati, and entrepreneurs in the twenty-first century enterprise / David J. Teece -- ch. 22. Looking to the future: bringing organizations deeper into human capital theory / Peter D. Sherer -- ch. 23. Human capital formation regimes: states, markets, and human capital in an era of globalization / Sean O Riain -- ch. 24. Human capital in developing countries: the significance of the Asian experience / Thomas Clarke -- ch. 25. The future of human capital: an employment relations perspective / Thomas A. Kochan and Adam Seth Litwin. Subjects: Human capital. Series: Oxford handbooks in business and management

The social dynamics of innovation networks / edited by Roel Rutten, Paul Benneworth, Dessy Irawati and Frans Boekema. LCCN: 2013048796 Published/Produced: New York: Routledge, 2014. Description: xxv, 284 pages; 24 cm. Links: Cover image http://images.tandf.co.uk/common/jackets/websmall/978041585/9780415859479.jpg ISBN: 9780415859479 (hardback) LC classification: HM1111 .S557 2014 Related names: Rutten, Roel, 1971- editor of compilation. Summary: "The social dynamics of innovation networks captures the important role of trust, social capital, institutions and norms and values in the creation of knowledge in innovation networks. In doing so, this book connects to a long-standing debate on the socio-spatial context of innovation in economic geography, which is usually referred to as the Territorial Models of Innovation (TIMs) literature. This present volume breaks with the TIM literature in several important ways. In the first place, this book emphasizes the role of individual agency because individuals and their networks are increasingly recognized as the principal agents of knowledge creation. Secondly, this volume looks at space as a continuous field of opportunity rather than as bounded territory with a set of endowments, such as knowledge base and social capital. Although individually these elements are not new to the TIM literature, it has thus far failed to grasp their critical implication for studying the social dynamics of innovation networks.

The approach to the socio-spatial context of innovation in this volume is summarized as Knowledge Economy 2.0. It emphasizes that human creativity is now the main source of economic value and that human creativity and knowledge creation is not an organized process within organizations, but happens bottom up in formal and informal professional and social networks of individuals that cut across multiple organizations. "-- Provided by publisher. Subjects: Social interaction--Economic aspects. Economic development--Technological innovations. Knowledge management--Social aspects. Business and Economics / General. Business and Economics / Development / Business Development. Business and Economics / Development / Sustainable Development. Series: Regions and cities

Training and practice for modern day archaeologists / John H. Jameson, James Eogan, editors. LCCN: 2012951306 Published/Created: New York: Springer, 2013. Description: xxiv, 277 p.: ill. (some col.); 24 cm. Links: Contributor biographical information http://www.loc.gov/catdir/enhancements/fy1311/2012951306-b.html Publisher description http://www.loc.gov/catdir/enhancements/fy1311/2012951306-d.html ISBN: 9781461455288 (acid-free paper) 1461455286 9781461455295 (e-book) LC classification: CC107 .T73 2013 Related names: Jameson, John H. Eogan, James. Contents: Archaeologists at the table: from community to global / John H. Jameson -- Discovering the archaeologists of Europe / Kenneth Aitchison -- Australian archaeology in profile: a survey of working archaeologists / Sean Ulm, Stephen Nichols, and Cameo Dalley -- No going back: remembering when British archaeology changed forever / Kenneth Aitchison -- From feast to famine?: the archaeological profession in Ireland in the new millennium / Patrizia La Piscopia, E. Eoin Sullivan, and Conor McDermott -- Competition and the development of authorised heritage discourses in a re-emergent Scottish nation / Malcolm A. Cooper -- Confirming relevance: how American and Canadian archaeologists are training youth and adults in archaeology, heritage studies, and community partnerships / Sherene Baugher -- Over qualified and under skilled?: training and professional development in the UK / Kate Geary -- Resource transformation: the history and status of the cultural resource management industry in the United States / Michael R. Polk -- Archaeology in the "real world": the training-practice disconnect in North American consulting archaeology / John P. McCarthy and Aaron Brummitt -- The emergence of the commercial sector in Irish archaeology 1987-2007: lessons to be learnt on research opportunities lost / Margaret Gowen -- Knowledge creation and publication in development-led archaeology in Ireland: some problems and possible solutions / Linzi Simpson -- Knowledge, value and the Celtic tiger / Brendon Wilkins -- The Malta Convention and contemporary Polish archaeology / Zbigniew Kobyliński -- Process and realisation of archaeological research / František Žák Matyasowszky -- The management of archaeology on national road projects in the Republic of Ireland / James Eogan -- Archaeology and development in a maritime context, highlighting Drogheda and the Boyne Estuary as a case study / Niall Brady and Edward Pollard. Subjects: Archaeology--Case studies. Archaeology--Vocational guidance. Notes: Includes bibliographical references and index. Series: One world archaeology One world archaeology (Springer (Firm))

Trialogus / Wyclif; edited and translated by Stephen E. Lahey. LCCN: 2012014619 Personal name: Wycliffe, John, -1384. Uniform title: Trialogus. English Published/Produced: Cambridge; New York: Cambridge University Press, 2013. Description: viii, 363 pages; 24 cm Links: Cover image http://assets.cambridge.org/97805218/69249/cover/9780521869249.jpg Contributor biographical information http://www.loc.gov/catdir/enhancements/fy1210/2012014619-b.html Publisher description http://www.loc.gov/catdir/enhancements/fy1210/2012014619-d.html Table of contents only http://www.loc.gov/catdir/enhancements/fy1210/2012014619-t.html ISBN: 9780521869249 LC classification: BR75 .W83513 2012 Related names: Lahey, Stephen E., translator. Summary: "John Wyclif is known for translating the Vulgate Bible into English, and for arguing for the royal divestment of the church, the reduction of papal power and the elimination of the friars and against the doctrine of transubstantiation. His thought catalyzed the Lollard movement in England and provided an ideology for the Hussite revolution in Bohemia. Wyclif's Trialogus discusses divine power and knowledge, creation, virtues and vices, the Incarnation, redemption and the sacraments. It consists of a three-way conversation, which Wyclif wrote to familiarize priests and layfolk with the complex issues underlying Christian doctrine, and begins with formal philosophical theology, which moves into moral theology, concluding with a searing critique of the fourteenth-century ecclesiastical status quo. Stephen Lahey provides a complete English translation of all four books, and the 'Supplement to the Trialogue', which will be a valuable resource for scholars and students currently relying on selective translated extracts"-- Provided by publisher. Contents: Machine generated contents note: Introduction; Book I; Book II; Book III; Book IV; Supplement to the Trialogus: On the Endowment of the Church. Subjects: Theology--Early works to 1800. Religion / Christian Church / History. Notes: Includes bibliographical references and index.

Understanding knowledge creation: intellectuals in academia, the public sphere and the arts / edited by Nikita Basov and Oleksandra Nenko. LCCN:

2014497150 Published/Created: Amsterdam; New York: Rodopi, ©2012. Description: xix, 208 pages: illustrations; 22 cm. ISBN: 9789042034624 (pbk.) 9042034629 (pbk.) LC classification: HM213 .U47 2012 Related names: Basov, Nikita. Nenko, Oleksandra. Summary: "Understanding Knowledge Creation: Intellectuals in Academia, the Public Sphere and the Arts brings together scholars from a wide range of disciplines and cultures and involves them into a multi-dimensional dialogue on the mechanisms of knowledge creation in the present-day society with a specific focus on intellectuals as knowledge creators in three main arenas of their activity: the 'institutionalized' arena - academia - and two adjacent arenas: the public sphere and the arts."--Publisher's website. Contents: Part 1: Intellectuals in Academia: Institutionalized Knowledge Creation. The Rationalization of Academia: From Bildung to Production --The Democratic Intellect Reconsidered. Part 2: Intellectuals in the Public Sphere: 'Organic' Knowledge Creation. Post-Apartheid Organic Intellectual and Knowledge Creation -- Walter A. Rodney and the Instrumentalist Construction and Utilization of Knowledge -- Possibilities and Risks of Influencing Public Knowledge: The Case of Hrant Dink -- Not a Sin, but a Side Effect: Collaboration and Knowledge Creation by the Organic Intellectuals. Part 3: Intellectuals in the Arts: Emotional Knowledge Creation. Art and the Passion of Intellect -- Emotional Intelligence: Literature, Ethics and Affective Cognition in J.M. Coetzee's Disgrace -- Aesthetic Emotional Experience: From Eye Irritation to Knowledge. Subjects: Knowledge management. Intellectual life--History--21st century. Education, Higher--International cooperation. Diffusion of innovations. Notes: "Volume 78, a volume in the Critical issues series 'Intellectuals, knowledge, power'"--Preliminary page. Includes bibliographical references. Series: At the interface/probing the boundaries, 1570-7113; 78 At the interface/probing the boundaries; v. 78.

Universities and the public sphere: knowledge creation and state building in the era of globalization / edited by Brian Pusser [and 3 others]. LCCN: 2011003453 Published/Created: New York: Routledge, Taylor and Francis Group, 2012. Description: xiii, 261 pages: illustrations; 24 cm. ISBN: 9780415878470 (alk. paper) 0415878470 (alk. paper) 9780203847848 (ebk) 0203847849 (ebk) 9780415824262 (pbk.) 0415824265 (pbk.) LC classification: LC67.6 .U57 2012 Related names: Pusser, Brian. Contents: Introduction and overview of the book / Brian Pusser [and others] -- The "public" contribution of universities in an increasingly global world / Simon Marginson -- Power and authority in the creation of a public sphere through higher education / Brian Pusser -- The global sorting machine: an examination of neo-racism among international students and postdocs / Jenny J. Lee and

Brendan Cantwell -- Reconsidering the realities of the international student market: a perspective from Japan and East Asia / Akiyoshi Yonezawa and Hugo Horta -- University rankings, global models, and emerging hegemony: critical analysis from Japan / Mayumi Ishikawa -- The corporatization of national universities in Finland / Jussi Välimaa -- Negotiating race and gender identity in the knowledge age: the case of South African universities / Reitumetse Obakeng Mabokela -- The broken promise of neoliberal restructuring of South African higher education / Ivor Baatjes, Carol Anne Spreen, and Salim Vally -- Higher education and the public sphere in Angola / Ken Kempner and Ana Jurema -- Selling out academia?: higher education, economic crises, and Freire's generative themes / Greg William Misiaszek, Lauren Ila Jones and Carlos Alberto Torres -- Post neoliberalism: the professional identity of faculty off the tenure-track / John S. Levin, Genevieve Shaker, and Richard Wagoner -- Chameleon's agenda: transformations of private universities within the framework of public higher education policies in Mexico / Roberto Rodriguez-Gomez and Imanol Ordorika. Subjects: Education, Higher--Economic aspects--Cross-cultural studies. Education and globalization--Cross-cultural studies. Privatization in education--Cross-cultural studies. Critical pedagogy. Neoliberalism. Critical pedagogy. Education and globalization. Education, Higher--Economic aspects. Neoliberalism. Privatization in education. Form/Genre: Cross-cultural studies. Notes: Includes bibliographical references and index. Series: International studies in higher education International studies in higher education.

Universities in the knowledge economy: higher education organisation and global change / edited by Paul Temple. LCCN: 2011027399 Published/Created: Milton Park, Abingdon, Oxon; New York: Routledge, 2012. Description: xiv, 250 pages: illustrations; 24 cm. ISBN: 9780415884662 (hardback) 0415884667 (hardback) 9780203142899 (e-book) 0203142896 (e-book) LC classification: LC67.6 .U578 2012 Related names: Temple, Paul, 1952- Summary: "Universities are fundamental to the contemporary knowledge economy. They directly and indirectly support economic growth in both developing and advanced economies. In addition to their traditional teaching and research functions, they often also have important roles in supporting regional development and urban regeneration, as well as involvement in fostering international relations, in, cultural developments and in enhancing social cohesion. While higher education institutions in many countries are often assigned key roles in economic and social policy prescriptions, exactly what those roles are and how they should be carried out are often unclear. Universities and the Knowledge Economy provides a much-needed theoretical and empirical analysis of these functions, taking a critical look at the complex connections between knowledge creation, the knowledge economy, and higher education today. This volume: Brings together work on these topics by international experts, reporting and analysing recent policy developments and research Shows the significance of the university's role in the knowledge economy, and the precise roles that it can play. Presents a range of studies showing how universities interact with other knowledge producers and users, and how these interactions can be managed to achieve the most effective applications of knowledge Universities are multi-faceted institutions that everywhere are accorded special status. Universities and the Knowledge Economy examines how these institutions carry our knowledge production and application, and how their distinctive characters affect what they do. This title is of both intellectual and operational relevance, and would be suitable for those interested in higher education and policy and practice, and in the theory of higher education. Paul Temple is Reader in Higher Education Management and Co-Director of the Centre for Higher Education Studies at the Institute of Education, University of London, UK"--Provided by publisher. Subjects: Education, Higher--Economic aspects. Economic development--Effect of education on. Education and globalization. Education--General. Education--Higher. Economic development--Effect of education on. Education and globalization. Education, Higher--Economic aspects. Hochschulbildung. Landesentwicklung. Regionale Wirtschaftsstruktur. Universität. Wirtschaftsentwicklung. Hochschule. Studium. Wirtschaftsentwicklung. Wissensproduktion. Globalisierung. Notes: Includes bibliographical references and index. Series: International studies in higher education International studies in higher education.

Visible past [electronic resource]. LCCN: 2011202238 Published/Created: Indianapolis, IN: Ideagora Publication history: Began with Mar. 1, 2010. Current frequency: Continuously updated ISSN: 2159-9378 LC classification: AZ195 Serial key title: Visible past Abbreviated title: Visible past Summary: A web-based knowledge creation and storage environment produced at Purdue University. Its primary aim is to help students and scholars to experience and communicate with fully immersive, historically accurate models of past geographic realities or to relate information to specific real geographic locations. It assists humanities and social science scholars in their daily work of organizing sources and writing up narratives. Subjects: Humanities--Electronic information resources--Periodicals. Electronic file info: http://www.visiblepast.net/see

RELATED NOVA PUBLICATIONS

KNOWLEDGE TRANSFER: PRACTICES, TYPES AND CHALLENGES

Editor: Dragan Ilic
Department of Epidemiology and Preventive Medicine,
School of Public Health and Preventive Medicine, Monash University,
Melbourne VIC, Australia
ISBN: 978-1-62081-538-0 Publication Date: 2012

The discipline of knowledge transfer has existed for many years across disciplines, but has more recently gained exposure within the health care field including medicine, nursing, health sciences, public health and health services research. Knowledge transfer is a complex iterative process that aims to implement high quality evidence into practice ensuring that better health outcomes are achieved across different levels including patients, health professionals, health systems and policy makers. This book provides an introduction to the field of knowledge transfer, examining the different approaches to knowledge transfer as well as detailing challenges to its implementation in practice. This book examines what knowledge transfer is, and how it may differ across different contexts and settings.

KNOWLEDGE TRANSFER IN A MULTIDISCIPLINARY HEALTH CARE ARENA: POSITIONS AND PLOTS

Editors: Carmen W.H. Chan
The Nethersole School of Nursing, The Chinese University of Hong Kong, Hong Kong, China
ISBN: 978-1-62081-160-3 Publication Date: 2012

This book is a collection of latest works that show the unique roles and positions of different health care professionals on the arena of health care knowledge transfer. Although the transfer of knowledge-based evidence into practice and service has long been advocated, the gap between the conduct of research and the application of research findings remains a concern. This book describes innovative knowledge transfer projects in real life situations, ranging from health care education, health promotion, disease prevention and control to end-of-life care.

KNOWLEDGE TRANSFER IN HEALTH CARE: AN OVERVIEW*

Chi Kwok Chan
Prince of Wales Hospital, Hong Kong,
Email: chansdu@hotmail.com

This chapter reviews the literature on the basics of knowledge management with particular focus on the healthcare industry. We start with the definition of knowledge and explain its difference from data and information before moving to the differentiation between tacit and explicit knowledge. The cycle of knowledge management is then briefly discussed. Following that, the concept of knowledge transfer is defined and its importance is expounded in

* The full version of this chapter can be found in *Knowledge Transfer in a Multidisciplinary Health Care Arena: Positions and Plots*, edited by Carmen W.H. Chan, published by Nova Science Publishers, Inc, New York, 2012.

details. This chapter lays the foundation for the subsequent chapters of the book.

HEALTHCARE PROJECT ON LOWER URINARY TRACT DYSFUNCTION (LUTS): THE IMPLEMENTATION OF A SHARED CARE MODEL IN KNOWLEDGE TRANSFER*

Chi Kwok Chan
Prince of Wales Hospital, Hong Kong, Sidney KH YIP,
Pedder Clinic, Hong Kong
Email: chansdu@hotmail.com

In this chapter the knowledge transfer (KT) process is in the context of transferring skills and capabilities, exchanging good ideas, research results, experiences and skills between individuals, universities, research organizations, government, the public sector and the wider community to enable innovative new products, services and policies to be developed with the main focus on healthcare professionals. The 5 elements (What? To whom? By whom? How? Evaluation?) and 3 models (Push / Pull / Exchange) of knowledge transfer are discussed in details with selected examples from a project about transfer of knowledge related to the clinical management of lower urinary tract symptoms (LUTS) from experienced workers and knowledge creators to the audience with different background exposure to this subject.

* The full version of this chapter can be found in *Knowledge Transfer in a Multidisciplinary Health Care Arena: Positions and Plots*, edited by Carmen W.H. Chan, published by Nova Science Publishers, Inc, New York, 2012.

PROMOTING END-OF-LIFE CARE AND ADVANCE DIRECTIVES IN HONG KONG: A KNOWLEDGE TRANSFER PROJECT FOR THE COMMUNITY AND CLINICIANS*

Carmen W. H. Chan[1,†], Michael M. K. Sham[2], Raymond S. K. Lo[3], Sek Ying Chair[1], Helen Y. L. Chan[1], David C. Y. Lai[1] and Ying Yu Chui[4]

[1]The Nethersole School of Nursing, The Chinese University of Hong Kong
[2]Grantham Hospital, Hong Kong
[3]Shatin Hospital and Bradbury Hospice, Hong Kong
[4]Affiliated with The Chinese University of Hong Kong, Hong Kong when the project was conducted

The goal of this knowledge transfer project was to promote public awareness and appreciation of the benefits of end-of-life care. Phase one of the project included 8 road-shows to disseminate the philosophy and advertise the services available as regards end-of-life care to the community in Hong Kong. Five of the road-shows were organized in hospital lobbies and three in shopping malls. The philosophy of end-of-life care, encompassing the idea of advance care planning and advance directive was disseminated. Participants were given sets of the promotional materials (pamphlet and VCD) and completed a brief survey. Phase two of the project involved a 2-day training workshop for 58 home care nurses and doctors on advance care planning and advance directives. The project has increased the community's awareness of, and interest in, the available services when a person is facing a life-limiting disease and encourage more openness in preparing for a good death. The project also provided updated training to nurses and doctors to improve delivery of end-of-life care.

---

* The full version of this chapter can be found in *Knowledge Transfer in a Multidisciplinary Health Care Arena: Positions and Plots*, edited by Carmen W.H. Chan, published by Nova Science Publishers, Inc, New York, 2012.

† Email of corresponding author: whchan@cuhk.edu.hk.

CHALLENGES OF KNOWLEDGE TRANSFER IN CLINICAL SETTINGS*

Dorothy N. S. Chan[1], Winnie P. Y. Tang[1], Winnie K. W. So[1,], Daniel Y. T. Fong[2]*

[1]The Nethersole School of Nursing, The Chinese University of Hong Kong

[2]School of Nursing, The University of Hong Kong

Knowledge transfer emphasizes on using systematic ways of capturing, collecting, sharing knowledge which can be implemented to user with positive outcomes. In clinical settings, transferring knowledge into health care services is essential so as to ensure quality of care and patient safety. To achieve these goals, development of evidence-based practice becomes more and more crucial in the process of transferring knowledge to daily practice. In this chapter, the authors will share their experience of how they initiated an evidence-based practice in both adult and neonatal units using the framework developed by Melnyk-Fineout-Overholt (2005). The process of its development will focus on problem identification, review and appraise evidence, integrate evidence and assessing challenges in practice decision.

---

* The full version of this chapter can be found in *Knowledge Transfer in a Multidisciplinary Health Care Arena: Positions and Plots*, edited by Carmen W.H. Chan, published by Nova Science Publishers, Inc, New York, 2012.

* Email of corresponding author WKW SO: winnieso@cuhk.edu.hk.

A New Era for Knowledge Transfer: Drug Information‡

Vivian W. Y. Lee#
School of Pharmacy, Faculty of Medicine,
The Chinese University of Hong Kong, Hong Kong

Drug-related questions are often encountered by people who are on medications or taking health supplements. However, due to various barriers, these questions remain unanswered leading to possible medication problems. The Outreach Programme organized by the School of Pharmacy at The Chinese University of Hong Kong showed that more than half (55.3%) of the drug-related problems discovered in the elderly population were associated with lack of knowledge about their medication and their state of disease. Another study released at the same year on high cholesterol patients also found that nearly 80% of the patients had inadequate knowledge on the possible side effects associated with their medication. Furthermore, sensitive medical issues, such as sexually transmitted diseases, erectile dysfunction, use of contraceptive etc., are not widely discussed among the public. In a study on the use of oral contraceptive pills (OCPs) conducted by us in 2008, more than half of the interviewees had misconceptions. With the increasing popularity of self-medication as the first-line treatment for mild diseases, together with the lack of drug knowledge, it is expected that more drug-related questions would arise in the future.

In response to the expected drug information demand, a new online interactive drug enquiry system called *AMPOULE* (*A*sk *M*y *P*harmacist! - *O*nline *U*niversity-*L*ed drug *E*nquiry) was established. Citizens in Hong Kong and other overseas enquirers can now pose any drug-related question through this convenient and innovative channel. Through *AMPOULE* (http://www.ampoule.org.hk.), pharmacists provide personalized answers to drug-related queries, hoping to reduce possible medication problems faced by the general public.

‡ The full version of this chapter can be found in *Knowledge Transfer in a Multidisciplinary Health Care Arena: Positions and Plots*, edited by Carmen W.H. Chan, published by Nova Science Publishers, Inc, New York, 2012.

Email: vivianlee@cuhk.edu.hk.

We target to enhance communication between pharmacists and the general public through innovative channels. The demand of drug information is expected to rise with more well-educated enquires. Enquirers expect fast and efficient professional assistance for their drug-related problems. The traditional pharmacist consultation in the pharmacy may not be sufficient to fulfill enquirers' demands. Innovative media is necessary to improve the delivery of drug information. It is hoped that the public awareness on proper medication use and safety can be enhanced by providing proper drug knowledge, thus improving the overall effectiveness of drug treatment. In this book chapter, the following objectives will be achieved:

1. Difference of various drug information delivery will be compared;
2. Factors affecting drug information delivery will be discussed;
3. Experiences in developing online drug information platform will be shared;
4. Address the needs of drug information vary in different patient populations.

THE USE OF GUIDED REFLECTION FOR CLINICAL LEARNING IN POST-GRADUATE MENTAL HEALTH NURSING STUDENTS: INSIGHTS INTO KNOWLEDGE TRANSFER*

Wai Tong Chien†
School of Nursing, Faculty of Health and Social Sciences
The Hong Kong Polytechnic University,
Hong Kong S.A.R., P. R. China

This study is designed to explore the usefulness of guided self-reflection in facilitating nursing students' learning in acute psychiatric care settings. During their one-month clinical placement, in a variety of acute in-patient psychiatric care settings, 28 first year mental health nursing students kept

* The full version of this chapter can be found in *Knowledge Transfer in a Multidisciplinary Health Care Arena: Positions and Plots*, edited by Carmen W.H. Chan, published by Nova Science Publishers, Inc, New York, 2012.

† Email:hschien@inet.polyu.edu.hk.

journals describing important daily events, reflecting on their attitudes towards the events and indicating what they could learn from them. The clinical practice coordinator (researcher) and two clinical mentors (psychiatric nurse specialists) supervised 5-6 students in groups, guided the development of their self-reflection, examined the content of their reflective journals and provided weekly feedback. All the students using guided reflection were able to apply a wide range of learned theories and the skills of mental health nursing to their self-reflection activities and clinical care. Major themes in the reflective journals coincided with the development of the students' self-reflection skills, including nurse-patient relationships, meaningful self-reflection processes and identification with patients. These issues, together with other areas of learning, were discussed and clarified with the students. The findings suggest that the guided reflective learning method can help clinical teachers and mentors to put knowledge and skills into practice and meet the learning needs of their students at the post-graduate level in various clinical areas of psychiatric care.

KNOWLEDGE TRANSFER ON CARDIOVASCULAR DISEASE PREVENTION AND MANAGEMENT*

Sek Ying Chair†
The Nethersole School of Nursing,
The Chinese University of Hong Kong, Hong Kong

Cardiovascular diseases including coronary heart disease, stroke and hypertension are the far leading causes of death in Hong Kong. With the increasing in aging population from 12.8% in 2009 to 26.4% in 2036, the prevalence of cardiovascular disease is expected to go even higher. However, many of the risk factors for cardiovascular diseases, such as tobacco use, physical inactivity, and an unhealthy diet, are modifiable and can be corrected by lifestyle modifications. Yet, successful lifestyle changes require significant commitment by the individual and the ongoing support from the healthcare

* The full version of this chapter can be found in *Knowledge Transfer in a Multidisciplinary Health Care Arena: Positions and Plots*, edited by Carmen W.H. Chan, published by Nova Science Publishers, Inc, New York, 2012.

† Email: sychair@cuhk.edu.hk.

team. Chinese materials of hypertension, coronary heart disease and stroke are found to be scattered and many are far difficult for a layman to understand. Thus, empowering the general public by using information technology support is essential to promote cardiovascular health, as well as to confine the healthcare cost of Hong Kong.

With the support of the Knowledge Transfer Project Fund from The Chinese University of Hong Kong, a new CVHealth website (www. cuhk.edu.hk/proj/CVHealth), is developed and launched for free public access. The website aims to empower the public to prevent and control the occurrence of coronary heart disease, stroke and hypertension. The website is supported by a multidisciplinary team of experts including cardiologists, pharmacologists, dieticians, physiotherapists, occupational therapists, medical social workers, as well as patient representatives. With the multiple local and national promotional strategies performed, this website has been well received by the Chinese living locally as well as living overseas.

In this chapter, a series of knowledge transfer measures for preventing and managing cardiovascular diseases will be presented. Though empirical cardiovascular research and knowledge have been well established, transferring the developed knowledge, especially culturally-relevant to Chinese, to empower public to fight against these diseases is limited.

HOW TO TRANSFER KNOWLEDGE TO THE COMMUNITY ON PROMOTING CARDIOVASCULAR HEALTH?*

Sek Ying Chair†
The Nethersole School of Nursing,
The Chinese University of Hong Kong, Hong Kong

Health compromising behaviours such as physical inactivity and poor nutrition are detrimental to human health, which finally leading to loss in efficiency and productivity of the individual and burdening the health care system. Eating healthily and integrating regular exercise in daily life is a

* The full version of this chapter can be found in *Knowledge Transfer in a Multidisciplinary Health Care Arena: Positions and Plots*, edited by Carmen W.H. Chan, published by Nova Science Publishers, Inc, New York, 2012.

† Email: sychair@cuhk.edu.hk

critical component in preventing heart disease as well as other chronic diseases such as stroke, hypertension and diabetes mellitus. By reducing the risk of getting these diseases in general public, the health care cost can definitely be confined especially in a government subsidized healthcare system like Hong Kong.

This project aimed to promote healthy behaviours of the general public regarding heart health favored lifestyle through the following activities:

- Community roadshows and health promotion exhibitions
- Dietary workshops and a 'HealthyMenu Design Competition'
- A Physical exercise program
- Production of an educational booklet.

A Happy Ending with HOPE: A Knowledge Transfer Empowerment Project on Advance Care Planning (ACP)*

Carmen W. H. Chan†
The Chinese University of Hong Kong, Hong Kong

Eighty-five percent of Hong Kong citizens have not heard of advance care planning (ACP) and advance directives (AD). An empowerment project, (HOPE), proposed in this chapter, consists of three key areas: (H) Health care professionals' empowerment, (O) Older people/family members' empowerment, and (PE) Public empowerment. Four multi-dimensional and interactive phases will be arranged to promote the awareness of the Hong Kong community about their right and autonomy of ACP and AD; to empower nurses and medical social workers (MSWs) with competence and confidence to provide one-on-one consultations on ACP and AD to older people and their families; and to engage older people (residential care home elders) and their family members in the discussion of end-of-life care and ultimately, to develop their living wills and AD documents.

* The full version of this chapter can be found in *Knowledge Transfer in a Multidisciplinary Health Care Arena: Positions and Plots*, edited by Carmen W.H. Chan, published by Nova Science Publishers, Inc, New York, 2012.

† Email: whchan@cuhk.edu.hk.

RESEARCH PARTNERSHIP AND KNOWLEDGE TRANSFER IN THE DEVELOPMENT OF A GENERIC EVALUATION TOOLKIT FOR HEALTH PROMOTION INTERVENTIONS IN PRIMARY CARE*

Aideen Dunne†*, Angela Scriven and Andrew Howe*

School of Health Sciences and Social Care, Brunel University, Uxbridge, Middlesex and Public Health Department, NHS Harrow, Middlesex, United Kingdom

Evaluation research is fundamental to generating evidence to inform the development of policy, services and practice in health promotion. At a national and international level, rigorous evaluation is central to advancing knowledge and practice. The speed at which health promotion practice and knowledge is advancing is potentially stalled by a number of complexities associated with the process of evaluating health promotion. This research aims to address some of these complexities of evaluating health promotion through the development of a Generic Evaluation Framework and Toolkit, which can be applied directly to practice. The research had two study groups; staff of the primary care organisation and members of the general public who accessed the services provided by the primary care organisation. A literature review and consultation with the research team and primary care staff informed the development of the Evaluation Framework and Toolkit. To test the reliability and validity of the Evaluation Toolkit, staff and service users of the primary care organisation were recruited. The final version of the Evaluation Toolkit will be developed as an electronic programme and further piloting with staff will be undertaken. The research collaboration between primary care staff and the University made important contributions to the development and progress of the research. The involvement of staff in research can also contribute to the process of translating evidence into decision making about future funding allocation, service planning and targeting local health needs.

---

* The full version of this chapter can be found in *Public Health Yearbook 2011*, edited by Joav Merrick, published by Nova Science Publishers, Inc, New York, 2012.

† Correspondence: Aideen Dunne BSc MSc, Research Associate, School of Health Sciences and Social Care, Mary Seacole Building, Brunel University, Uxbridge, Middlesex, UB8 3PH United Kingdom. Email: aideen.dunne@brunel.ac.uk

WHAT IS KNOWLEDGE TRANSFER?*

Dragan Ilic†
Department of Epidemiology and Preventive Medicine,
School of Public Health and Preventive Medicine,
Monash University, VIC, Australia

Knowledge transfer aims to close the gap between evidence and practice. It aims to identify and implement knowledge into practice in a timely manner to ensure that optimal outcomes are met.

The methods to close the 'evidence to practice' gap form the framework for knowledge transfer. In order to bridge this gap, acknowledge transfer framework needs to identify information gaps, create new knowledge (or obtain existing knowledge), effectively implement it into the relevant setting and evaluate its effectiveness.

This chapter will provide a commentary defining the concept of knowledge transfer and outline the development of the DIER knowledge transfer framework.

---

* The full version of this chapter can be found in *Knowledge Transfer: Practices, Types and Challenges*, edited by Dragan Ilic, published by Nova Science Publishers, Inc, New York, 2012.

† Corresponding author: E-mail: dragan.ilic@monash.edu

Knowledge Transfer and Organizational Learning through Strategic Alliances: Theoretical Review and Research Perspectives[‡]

Ana Aleksić Mirić[#]
Faculty of Economics, University of Belgrade, Serbia

Alliances as a form of inter-organizational cooperation are increasing in number and importance on the global business scale, and numerous reasons can be given to explain this. Among other benefits that alliances can bring to those who create them, organizational learning and knowledge transfer have become particularly important. Alliances are used as a means of achieving both single and double loop learning, for exploration as well as for the exploitation purposes, and are thought to be particularly powerful in tacit knowledge transfer. Theoretical review in this chapter addresses: first, the issues of knowledge transfer and organizational learning in strategic alliances, next, it identifies potential research gaps and finally, it gives propositions that might help future research in this field.

[‡] The full version of this chapter can be found in *Knowledge Transfer: Practices, Types and Challenges*, edited by Dragan Ilic, published by Nova Science Publishers, Inc, New York, 2012.

[#] E-mail address: anaa@ekof.bg.ac.rs

Challenges and Opportunities in Knowledge Transfer by Informal and Formal Knowledge Brokers that Support Persons with Chronic Pain*

Lynn Shaw[1,†], *Joy MacDermid*[2], *Lubna Daraz*[3] *and Bill Chedore*

[1]School of Occupational Therapy, Faculty of Health Sciences, The University of Western Ontario, London, Ontario Canada

[2]School of Rehabilitation Science, McMaster University, Hamilton, Ontario, Canada

[3] Dept.of Psychiatry and Behavioural Neuroscience. McMaster University, Hamilton, Ontario, Canada

Knowledge brokers such as health care professionals and consumer group peer support leaders experience many challenges in transferring and exchanging knowledge for persons with chronic pain and those with work injuries. Meeting the needs of end users and their preferences for considering and integrating knowledge into decision making is not easy. This chapter will provide an overview of the barriers to KT and offer strategies to facilitate ways to meet the needs of end users such as persons with chronic pain that is also sensitive to the contextual and social cultural expectations of the end users. Readers will be introduced to an Interactive Knowledge Transfer Process that can enable formal and informal knowledge brokers to create spaces and opportunities for practical knowledge exchange of research, procedural and health information needed to manage daily and work activities while living with pain. This chapter will also have a focus on what knowledge brokers need to know to help end users successfully appraise information available on the internet.

---

* The full version of this chapter can be found in *Knowledge Transfer: Practices, Types and Challenges*, edited by Dragan Ilic, published by Nova Science Publishers, Inc, New York, 2012.

† Corresponding author: E-mail: leshaw@uwo.ca

Utilising Evidence Based Practice to Facilitate Knowledge Transfer*

***Dragan Ilic*†**
Department of Epidemiology and Preventive Medicine,
School of Public Health and Preventive Medicine,
Monash University, VIC, Australia

The principles of evidence based practice focus on providing users with the skills to identify, evaluate and implement evidence within a clinical setting. A similar process is also apparent with knowledge transfer, which also takes into account barriers and facilitators of evidence uptake by stakeholders. This chapter will explore how the principles of evidence based practice may be utilised by individuals to facilitate knowledge transfer from a user's perspective.

Parametric Link Models for Knowledge Transfer in Statistical Learning*

F. Beninel[1], C. Biernacki[2], C. Bouveyron[3], J. Jacques[2] and A. Lourme[4]
[1]CREST-ENSAI, Bruz, France
[2]Université Lille 1 and CNRS and INRIA, Lille, France
[3]Université Paris 1 Panthéon-Sorbonne, Paris, France
[4]Université de Pau et des Pays de l'Adour, Pau, France

When a statistical model is designed in a prediction purpose, a major assumption is the absence of evolution in the modeled phenomenon between

* The full version of this chapter can be found in *Knowledge Transfer: Practices, Types and Challenges*, edited by Dragan Ilic, published by Nova Science Publishers, Inc, New York, 2012.

† Corresponding author: E-mail: dragan.ilic@monash.edu

* The full version of this chapter can be found in *Knowledge Transfer: Practices, Types and Challenges*, edited by Dragan Ilic, published by Nova Science Publishers, Inc, New York, 2012.

the training and the prediction stages. Thus, training and future data must be in the same feature space and must have the same distribution. Unfortunately, this assumption turns out to be often false in real-world applications. For instance, biological motivations could lead to classify individuals from a given species when only individuals from another species are available for training. In regression, we would sometimes use a predictive model for data having not exactly the same distribution that the training data used for estimating the model. This chapter presents techniques for transferring a statistical model estimated from a *source* population to a *target* population. Three tasks of statistical learning are considered: Probabilistic classification (parametric and semi-parametric), linear regression (including mixture of regressions) and model-based clustering (Gaussian and Student). In each situation, the knowledge transfer is carried out by introducing parametric links between both populations. The use of such transfer techniques would improve the performance of learning by avoiding much expensive data labeling efforts.

PATTERNS OF KNOWLEDGE TRANSFER IN THE SPANISH BIOTECHNOLOGY INDUSTRY*

Beatriz Ortiz[1], Mario J. Donate[2,†] and Fátima Guadamillas[3]

[1]University of Castilla-La Mancha, Faculty of Law and Social Sciences, Department of Business Administration, Cobertizo San Pedro Mártir s/n, Toledo, Spain

[2]University of Castilla-La Mancha, Faculty of Law and Social Sciences, Department of Business Administration, Ronda de Toledo s/n, Ciudad Real, Spain

[3]University of Castilla-La Mancha, Faculty of Law and Social Sciences, Department of Business Administration, Cobertizo San Pedro Mártir s/n, Toledo, Spain

---

* The full version of this chapter can be found in *Knowledge Transfer: Practices, Types and Challenges*, edited by Dragan Ilic, published by Nova Science Publishers, Inc, New York, 2012.

† Corresponding author: E-mail: mariojavier.donate@uclm.es

The biotechnology industry is characterized by a high intensity in knowledge creation and innovation. Biotechnology companies are constantly affected by knowledge flows so this industry is an interesting field to study knowledge transfer strategies. In this chapter we analyze the knowledge transfer strategies of three Spanish biotechnology companies, PharmaMar, Neocodex and Digna Biotech. With this aim an exploratory case study was carried out by the authors, which is based on the model of technological knowledge transfer developed by Bozeman (2000). The results of this study provide a comprehensive overview of the aspects that characterize different ways to understand a knowledge transfer strategy and the results that are obtained from its application.

Academic and Scientific Poster Presentation - Visualising Knowledge as a Means to Facilitate Knowledge Transfer*

***Nicholas E. Rowe*†**

Freelance Academic Writer and Editor - Pello, Finland

Poster presentations are a common-place feature of many modern academic and professional settings. Based upon the wide range of professional groups that utilise the medium and the multitude of local, national and international meetings at which they are presented, it may be fair to estimate that millions are undertaken every year. As a stand-alone medium however, they are not best suited to facilitate knowledge transfer. This chapter looks at how we can make the most of the many opportunities for poster presentation and use facilitated dialogue to increase the potential to communicate our findings, opinions and research to others. In particular, it considers those aspects of knowledge transfer commonly found in both the business and conference setting; namely the concepts of knowledge / research dissemination and its subsequent utilization by the receiver. In considering these aspects, this chapter looks to demonstrate that although the academic poster is a favoured

* The full version of this chapter can be found in *Knowledge Transfer: Practices, Types and Challenges*, edited by Dragan Ilic, published by Nova Science Publishers, Inc, New York, 2012.

† E-mail address: editorinchief@thewrittenword.eu

medium of visualising knowledge, there are further considerations to be made, if we are to effectively transfer that knowledge to others.

KNOWLEDGE TRANSFER AND PRACTICE FOR CLINICAL SKILLS PERFORMANCE*

***Stephen Maloney*†**
Department of Physiotherapy, School of Primary Health,
Monash University, Australia

Effective knowledge transfer between researchers, educators, and clinicians is essential for maintaining best practice in the health disciplines. It is the method by which we aim to enhance the health service delivery for our communities. This chapter examines methods of knowledge transfer for supporting health professional clinical performance.

Competent clinical skills are a combination of theoretical knowledge, manual handling, and effective communication, encapsulated in the clinician-client interaction. Training for optimum performance of clinical skills is a complex and enduring task. Health professional education is a continuum, beginning from entry into an undergraduate program, and continues until the health professional leaves their chosen discipline. This chapter focuses on innovative methods utilised within an undergraduate clinical skills program, techniques for supporting transfer of knowledge to a clinical environment, and the continuing education of health professionals. It tackles issues relating to access and utilisation of educational technologies, and facilitation of a successful knowledge transfer culture.

* The full version of this chapter can be found in *Knowledge Transfer: Practices, Types and Challenges*, edited by Dragan Ilic, published by Nova Science Publishers, Inc, New York, 2012.

† Corresponding author: E-mail: stephen.maloney@monash.edu

WHAT ROLE CAN JOURNAL CLUBS PLAY IN KNOWLEDGE TRANSFER?*

***Dragan Ilic*†**

Department of Epidemiology and Preventive Medicine,
School of Public Health and Preventive Medicine,
Monash University, VIC, Australia

Journal clubs have been operating in various formats for over 100 years in the medical field. Journal clubs aim to promote effective knowledge transfer of medical evidence to clinicians and subsequent uptake into the clinical setting.

Clinician awareness, agreement, adoption and adherence of evidence may all act as barriers when translating high quality evidence into practice.

This chapter will provide a commentary on the use of journal clubs in knowledge transfer and also proposes a method of utilizing journal clubs as complex interventions as a strategy to translate evidence into practice effectively.

* The full version of this chapter can be found in *Knowledge Transfer: Practices, Types and Challenges*, edited by Dragan Ilic, published by Nova Science Publishers, Inc, New York, 2012.

† Corresponding author: E-mail: dragan.ilic@monash.edu

IMPROVING COASTAL KNOWLEDGE TRANSFER BETWEEN RESEARCHERS AND MANAGERS: A TWO-WAY ROUTE*

M. M. Carapuço[1,*•], R. Taborda[1,†], C. Andrade[2,‡], and M. C. Freitas[2,§]

[1]Instituto Dom Luiz, University of Lisbon,
Campo Grande, Lisbon, Portugal

[3]Centro de Geologia da Universidade de
Lisboa, University of Lisbon,
Campo Grande, Lisbon, Portugal

It is well recognized that the scientific knowledge on coastal dynamics should contribute to coastal zone management. However, so far, generation of data on coastal processes have been conducted almost exclusively by the research community; the data sets and results are made available in ways that suit that community, but frequently preclude direct application in coastal management. Moreover, there is a plethora of site-specific information that is frequently detained by manager, which is often overlooked by the research community. This chapter assesses both managers and researchers perspective on this subject. Results show that the major gap in knowledge transfer is related with the channels used for data dissemination, which are not adequate to reach managers. This stresses the importance of fostering the existing mechanisms to improve coastal knowledge transfer such as the use of coastal indicators, the existence of linkage agents, the development of "qualitative into

---

* The full version of this chapter can be found in *Beaches: Erosion, Management Practices and Environmental Implications*, edited by Margaret Cessa, published by Nova Science Publishers, Inc, New York, 2014.

• *Corresponding author: M. M. Carapuço. Instituto Dom Luiz, University of Lisbon, Campo Grande, 1749-016 Lisbon, Portugal, Phone: +351 217 500 376, Fax: +351 217 500 065, E-mail: mmcarapuco@fc.ul.pt.

† R. Taborda: Instituto Dom Luiz and Departamento de Geologia, Faculty of Sciences, University of Lisbon, Campo Grande, 1749-016 Lisbon, Portugal, Phone: +351 217 500 376, Fax: +351 217 500 065, E-mail: rtaborda@fc.ul.pt.

‡ C. Andrade: Centro de Geologia da Universidade de Lisboa and Departamento de Geologia, Faculty of Sciences, University of Lisbon, Campo Grande, 1749-016 Lisbon, Portugal, Phone: +351 217 500 376, Fax: +351 217 500 065, E-mail: candrade@fc.ul.pt.

§ M. C. Freitas: Centro de Geologia da Universidade de Lisboa and Departamento de Geologia, Faculty of Sciences, University of Lisbon, Campo Grande, 1749-016 Lisbon, Portugal, Phone: +351 217 500 376, Fax: +351 217 500 065, E-mail: cfreitas@fc.ul.pt.

quantitative" data tools and the development of managers-oriented tools to support self-data acquisition. Additionally, further meetings and seminars involving researchers and managers - "get together" mechanics - are also suggested by both parties. The perception of coastal researchers and managers concerning gaps in knowledge transfer reflects an unexpected general agreement; this proximity can constitute a major opportunity in bridging these communities together and in the implementation of a two-way route communication approach, thus increasing and improving the role of managers in the coastal knowledge transfer processes.

KNOWLEDGE CREATION AND DIGITAL COLLABORATION IN HIGHER EDUCATION*

Soraya García-Sánchez•
Department of Modern Languages,
Universidad de Las Palmas de Gran Canaria, Spain

This chapter establishes a debate about the concepts of creativity and collaboration with regards to learning English as a Second Language in Higher Education. The chapter focuses on a heuristic, mindful learning method set in a blended learning environment of English for Specific Purposes (ESP) at the Universidad de Las Palmas de Gran Canaria, Spain. Various task-based examples are provided to demonstrate that a mindful, collaborative and creative learning approach that allows for reflection, participation and autonomy can contribute to enhance ESP competences effectively. On the one hand, this chapter analyses the different communicative tasks that enabled learners to be collaborative and creative in class and online. On the other hand, the results will corroborate that active participants of their learning process create knowledge that is closed to their own spaces. Creation and collaboration for ESP students have benefits that contribute to a successful communicative approach of foreign language and ITC competences.

---

* The full version of this chapter can be found in *Collaborative Learning: Theory, Strategies and Educational Benefits*, edited by Stephen Rutherford, published by Nova Science Publishers, Inc, New York, 2014.

• soraya.garcia@ulpgc.es

KNOWLEDGE TRANSFER IN COMMUNITY-EMBEDDED LEARNING: A CASE STUDY*

Wendy Fasso and Bruce Allen Knight
Central Queensland University, Australia

A design approach to the distance delivery of professional development in the area of e-learning design and facilitation was taken. The intent of the design was to address the problem of the transfer of learning in a distance context, modelled on Kazmer's (2007) community-embedded learning framework. Strategies were designed to support the two-way transfer of knowledge between the online social world and the workplace community. The process of transfer was identified as being through the generalization of specific experiences and knowledge, and its subsequent application to the local context, or to the online social community. The process of generalization was regarded as being critical to the transfer of knowledge. Data obtained through an analysis of student materials provided indications that generalization and transfer of knowledge had occurred within the diverse range of teaching contexts of participants and the online social environment. The result is a series of design principles that may well support the design of community embedded learning in other contexts.

* The full version of this chapter can be found in *Progress in Education. Volume 34*, edited by Roberta V. Nata, published by Nova Science Publishers, Inc, New York, 2015.

International Knowledge Transfer and Management: The Case of Kodak-ITRI Collaboration*

***Frank J. W. Shiu*[1], *Connie Zheng*[2,•] *and Mei-chih Hu*[3]**
[1]Industry Technology Research Institute (ITRI), Taiwan
[2]Deakin University, Australia
[3]National Tsing-hua University, Taiwan

This chapter uses the theoretical framework of strategic alliance learning (Siminon, 2004) and evaluates the complex process of international knowledge transfer and management between Kodak (USA) and the Industry Technology Research Institute (ITRI) of Taiwan. By analyzing the historical events and internal organizational routines and mechanisms of ITRI, it is found that successful international knowledge management of technology transfer is determined by several strategic learning variables such as learning intent, learning capacity, knowledge complementarities, and strategic management of tacit knowledge, apart from organizational size, structure and culture. Selection of key staff with language and cultural competencies is imperative in managing the success of international knowledge and technology transfer. In addition, top management commitment with low knowledge protectiveness from the teaching partner (Kodak) is also regarded as critical to facilitate the effective learning by the receiving partner (ITRI). Implications to international human resource management policies and practices are discussed.

* The full version of this chapter can be found in *International Human Resource Management: Trends, Practices and Future Directions*, edited by Connie Zheng, published by Nova Science Publishers, Inc, New York, 2016.

• Corresponding author, Deakin University, Deakin Business School, Department of Management, Faculty of Business and Law, 70 Elgar Road, Burwood, Victoria 3125, Australia. Email: connie.zheng@deakin.edu.au.

INDEX

A

B

C

D

E

F

L

M

N

O

P

R

S

T

U

V

W